LATE PLEISTOCENE VERTEBRATE PALEOECOLOGY
OF THE WEST

Late Pleistocene Vertebrate Paleoecology of the West ARTHUR H. HARRIS

 University of Texas Press, Austin

International Standard Book Number 0-292-74645-8
Library of Congress Catalog Card Number 84-51374
Requests for permission to reproduce material from this work
should be sent to Permissions, University of Texas Press, Box
7819, Austin, Texas 78713.

For reasons of economy and speed this volume has been
printed from camera-ready copy furnished by the author, who
assumes full responsibility for its contents.

CONTENTS

ACKNOWLEDGMENTS

Numerous people have aided this study in one way or another and are extended my sincere thanks. William B. Gillespie, George T. Jefferson, Lloyd E. Logan, R. Lee Lyman, Henry J. Messing, and Robert Reynolds have generously supplied unpublished data. Material from the collections of the Los Angeles County Museum of Natural History, the Museum of Southwestern Biology, the National Museum of Natural History, the Philadelphia Academy of Science, and the Texas Memorial Museum has been inspected, either by loan or visitation, thanks to the kindness of their respective curators. As always in the case of works of this nature, numerous colleagues have helped shape my concepts through the years by discussion and criticism. My final, but not least, thanks go to Celinda R. Crews, whose able help alleviated much of the drugery involved in manuscript preparation, editing, and proofing.

LATE PLEISTOCENE VERTEBRATE PALEOECOLOGY OF THE WEST

I. INTRODUCTION

This is a study of paleoecology--of relationships between organisms and their environments during an earlier time. The time in this case is the last part of the Pleistocene Epoch, that span of time running from about 1.8 million years ago to about 10,000 years before present (BP). More specifically, this book will be considering the last 120,000 years or so of the Pleistocene. This stretch of time includes the last major interglacial (the Sangamonian) and the last major glacial (the Wisconsinan). This period of time is of particular interest because the events of the Pleistocene led to the conditions under which man now lives, and modern environmental biology (including that of man) is hardly understandable without knowledge of these events.

The Pleistocene was a time of greatly fluctuating climatic conditions. Although often thought of as the Ice Age, the great continental and mountain glaciers were but one manifestation of the Pleistocene climates: changes in temperature regimes, moisture relationships, air movements, and sea levels occurred world-wide and caused concomitant changes in plant and animal distributions and directions of evolution. Moreover, within the Pleistocene were interglacials which brought conditions similar to those of today or possibly to degrees even less glacial. The glacial times themselves were not isoclimatic, but times of widely (perhaps wildly) fluctuating conditions. Neither were geologic conditions stable;

much of the surficial geology was directly affected by the climate, but tectonic movements, volcanism, and the like also played a significant part in the shaping of the Pleistocene landscape.

Although the time span considered here is brief compared to the Pleistocene as a whole, it does include examples of most of the conditions known from the entire epoch, and thus can be taken as broadly representative of the Pleistocene. From the point of view of reconstruction of the paleoecology, the recency of the geologic evidence and the occurrence of many organisms whose little-changed descendents are still extant allow reconstructions of climatic, geologic, and biologic events that would be impossible for the earlier Pleistocene but which allow, in a general way, the application of the findings to earlier times.

The region covered by this study is the western United States and adjacent areas of Canada and Mexico (Fig. 1). For purposes of organizing data, the eastern boundaries of Montana, Wyoming, Colorado, and New Mexico, along with Trans-Pecos Texas, mark the eastern limits of the study area; in Canada the area includes Alberta, British Columbia, and Saskatchewan. In Mexico, the border states of Baja California Norte, Sonora, and Chihuahua are considered, but known sites are scarce. I have not hesitated to violate these geographic guidelines when something of value could be gained.

The organisms of focus are the terrestrial mammals and a few terrestrial birds. There are a number of reasons for choosing a limited number of organisms for study. Not the least of these is the impractibility of doing justice to a more inclusive group. Beyond this, however, are other factors: the present distributions, systematics, and biologies

of the higher vertebrates are far better known than those of
most other animals; generally, fossil preservation, numbers
of specimens in collections, and ubiquity of distribution is
superior to those of other terrestrial organisms; the biology
of birds and mammals (warmblooded and with one kind or another adapted to each of a very wide variety of environments)
generally makes them more suited for study; and not least, my
knowledge of birds and mammals is greater than for other
organisms and I necessarily have to work within my own limitations.

This should not be taken to mean that evidence from other
fields will not be considered. Animal ecologists, neo- or
paleo-, can ignore neither the physical realities of the
environment nor the biotic features outside of their particular area of interest. Thus botanic and geologic evidence
will be utilized in this study, as well as pertinent information from other groups.

Of the two groups of organisms concentrated upon here,
birds are of considerable value in determining such things as
presence or absence of various habitat types (such as marsh,
open water, forest, etc.), but offer less than the mammals in
general when it comes to more specific details. This is due
in part to flight. Most birds readily pass from one suitable
habitat to another over unsuitable regions; a certain percentage invariably succumbs to illness, predation, or inclement
weather conditions enroute, their remains thus ending up in
places that the living animal would find insupportable.

Also important are the dispersal events common to most
animals. In birds, flight allows for a much wider dispersion
than is common among non-volant forms.

But perhaps most important are the seasonal migrations so
common among birds. During migration, individuals briefly

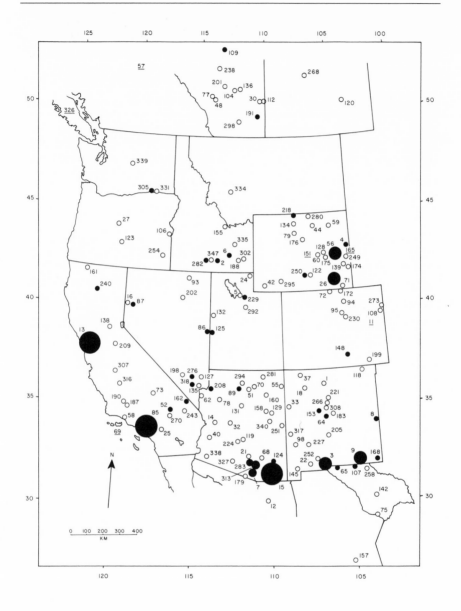

Fig. 1. Study area, showing fossil sites (some sites are outside the map area and mappable positions for some are unknown). Numbers refer to key below and to site numbers in Appendix 3. Solid symbols on map indicate multiple sites; numbers in caption that refer to these dots are in boldface, as are numbers referring to general areas (e.g., a county). A circle shows the approximate position of a single site. Sites shown are: 1. Abiquiu 2. Acequia; Minidoka; Rupert 3. Aden Fumerole; Anthony Cave; Conkling Cavern; Cueva Las Cruces; Picacho; Shelter Cave 4. Agate Basin; Brewster Site; Sheaman Site 5. Alpine Formation 6. American Falls; Dam; Duck Point; Rainbow Beach 7. Anaconda Pit; Calera Ranch; Canez Wash: No. 1, No. 2; Cerro Colorado: No. 1, No. 2; Cerros Negros 8. Anderson Basin/Circus Basin/Elephant Tusk Basin; Blackwater Draw; Blackwater Loc. No. 1--Brown Sand Wedge; Blackwater Draw No. 1--Gray Sand; McCullum Ranch 9. Animal Fair; Bison Chamber; Burnet Cave; Camel Room; Dark Canyon Cave; Harris' Pocket; Hermit's Cave; Human Corridor; Lost Valley; Muskox Cave; New Cave; Pit N & W Animal Fair; Room of the Vanishing Floor; Sabertooth Camel Maze; Slaughter Canyon; South Chimney; Stalag 17; TT II 11. Arikaree River 12. Arizpe 13. Arroyo Las Positas; Burke Ranch; California Sand & Gravel Co. Pit; Doolan Canyon; Livermore; Rodeo; San Francisco Bay Area 14. Artillery Mts. 15. Ash Canyon, Big Tooth; Boquillas Station; Brophy Cienega; Choate Ranch; Donnet; Double Adobe; Douglas; Dragoon Mts.; Elgin School; Empire South; Escapule; Fenn Site; Gray Site; Hereford Dairy; Horsethief Draw; Hurley; Lehner Ranch; Leikum; Lewis Hill; Lindsey Ranch; Mosan Ranch; Murray Springs: Arroyo Fauna, MS Conduit, Occupation Level, Portell Conduit, Unit D, Unit E; Naco; Papago Springs Cave; Pirtleville; Pomerene West; Pyeatt Cave; San Pedro Valley; San Rafael Aq.; Schaldack; Seff Loc.; South of Charleston; Tombstone Gulch; Wakefield; Whitewater Draw 16. Astor Pass 18. Atlatl Cave 21. Bain Site 22. Baldy Peak Cave 24. Bear Lake Region 25. Beaumont 26. Bell Cave; Bosler Gravel Pit; Dover; Horned Owl Cave; Laramie; Monolith Gravel Quarry 27. Bend 30. Bindloss 32. Black Mtn. 33. Black Rocks 37. Bloomfield 40. Brass Cap Pt. 42. Bridger 44. Buffalo 48. Calgary 51. Cameron 52. Camp Cady; Manix Lake 55. Canyon De Chelly 56. Careyhurst; Converse County; Douglas; Shawnee Creek 57. Cariboo District 58. Carpinteria 59. Carter/Kerr-McGee Site 60. Casper 62. Catclaw Cave 64. Cave in Manzano Mts.; Manzano Cave; Quarai 65. Ceremonial Cave; Hueco Tanks No. 1; Navar Ranch No. 13; Tank Trap Wash 69. Channel Islands 70. Charley Day Spr. 71. Cheyenne 72. Chimney Rock Animal Trap 73. China Lake 75. Cinnabar Mine 77. Cochrane 78. Coconino Cavern 79. Colby Site/Worland Area 85. Costeau Pit; La Mirada; Los

Angeles Basin; Newport Bay Mesa: Loc. 1066, Loc. 1067; Rancho La Brea; San Pedro; Zuma Creek **86.** Council Hall Cave; Smith Creek Cave; Smith Creek Cave: No. 4, No. 5; Streamview: No. 1, No. 2 **87.** Crypt Cave; Fishbone Cave **89.** Cylinder Cave; Tooth Cave; Tse'an Bida; Tse'an Kaetan; Tse-an Olje **93.** Deer Creek Cave **94.** Dent **95.** Denver **98.** Doolittle Cave **104.** Drumheller **106.** Dunniway Gravel Pit **107.** Dust Cave; Lower Sloth Cave; Upper Sloth Cave; Williams Cave **108.** Dutton **109.** Edmonton Area; Fort Saskatchewan **112.** Empress **118.** Folsom Site **119.** Fort McDowell **120.** Fort Qu'Appelle **122.** Fort Steel Site **123.** Fossil Lake **124.** Gardner Gravel; Wilcox Gravel **125.** Garrison: No. 1, No. 2 **127.** Glendale **128.** Glenrock **129.** Goodwater Wash **131.** Government Cave **132.** Granite Canyon No. 1 **134.** Greybull **135.** Gypsum Cave **136.** Hand Hills **138.** Hawver Cave **139.** Hell Gap Site **142.** Hord Rock Shelter **145.** Howell's Ridge Cave **148.** Huerfano Co.; Walsenburg **151.** Independence Rock Gravel Pit **153.** Isleta Caves: No. 1, No. 2 **155.** Jaguar Cave **157.** Jimenez Cave **158.** Joseph City **160.** Keams Canyon **161.** Klamath River **162.** Kokoweef Cave; Mescal Cave **165.** Lance Creek **168.** Lea Co.; San Simon Sink **172.** Lindenmeier **174.** Lingle **175.** Little Box Elder Cave **176.** Little Canyon Creek **179.** Lopez **183.** Lucy Site **187.** Maricopa **188.** McCammon **190.** McKittrick **191.** Medicine Hat: Fauna 2, Fauna 3, Fauna 4, Fauna 5, Fauna 6, Fauna 7 **198.** Mercury Ridge **199.** Mesa De Maya **201.** Milan Site **202.** Mineral Hill Cave **205.** Mockingbird Gap **208.** Muav Caves; Rampart Cave; Vulture Cave **209.** Murphreys **218.** Natural Trap Cave; Prospects Shelter **221.** New La Bajada Hill **224.** Nichols Site **227.** Palomas Creek Cave **229.** Park City; Silver Creek **230.** Parker **238.** Ponoka **240.** Potter Creek Cave; Samwel Cave **243.** Providence Mts. **249.** Rawhide Butte Area **250.** Rawlins; Union Pacific Mammoth Site **251.** Richville Gravels **252.** Robledo Cave **254.** Rome **258.** Salt Creek **266.** Sandia Cave **268.** Saskatoon **270.** Schuiling Cave **273.** Selby **276.** Sheep Range: Flaherty Mesa No. 1, Long Canyon Saddle Midden, Southcrest Midden **280.** Sheridan Area **281.** Shonto **282.** Shoshone Falls; Twin Falls **283.** Silver Bell Mts.; Wolcott: No. 2, No. 4, No. 5 **292.** Springville **294.** Stanton Cave **295.** Stauffer Chemical Plant **298.** Taber **302.** Thatcher Basin **305.** Touchet; Touchet Beds **307.** Tranquility **308.** Tree Springs **313.** Tucson Brickyard; Tucson Mts.; Tucson Mts. No. 1 **316.** Tulare Lake **317.** Tularosa Cave **318.** Tule Springs: Unit B-2, Unit D, Unit E-1 **326.** Vancouver Is. **327.** Ventana Cave **331.** Walla Walla **334.** Warm Springs No. 1 **335.** Wasden **338.** Wellton Hills **339.** Wenatchee **340.** Whipple Gravels **347.** Wilson Butte Cave

utilize habitats unsuitable for long-term occupation, move
through geographic areas where they otherwise are unknown,
and are particularly susceptible to the hazards mentioned
earlier. Moreover, because of the migratory habit, many
areas are utilized for weeks or months that could not support
either the numbers or kinds on a year-around basis.

These drawbacks to paleoecological interpretation, as well
as those held in common with many mammals (such as the wide
ecological tolerances of many predators), mean that relatively few kinds of birds are of prime importance in interpretation in our area, and only a few are utilized.

Although numerous mammals have wide environmental tolerances, many are tied to rather stringent limits today. These
are mostly small forms of limited vagility and often are
identified to species only with difficulty even from well-
preserved, plentiful material. The ecologically limited
forms are more numerous than in birds and most interpretation is based on the mammalian fauna.

The general organization is to present a brief introduction to paleoecologic theory as applied to Late Pleistocene
vertebrates, followed by summaries of the present pertinent
geography, climate, and biology of the region as a baseline.
A sketch of the time framework follows.

A brief introduction to the Late Pleistocene birds and
mammals is given, followed by interpretation of the paleontologic record from selected late Wisconsinan full-glacial
and Sangamonian interglacial sites. Interstadial sites are
considered next, and then a series of sites from a single
locality, varying in age from interstadial to Holocene, is
investigated.

Appendix 1 includes data concerning living taxa of importance to the interpretation, along with comments on each

extinct form. The fossil sites and their contained higher vertebrate faunas form the primary data base for the study of past environments. This full data base is given in Appendices 2 (taxa with sites of occurrence) and 3 (sites, with taxa occurring in each).

II. PALEOECOLOGICAL THEORY

A brief consideration of paleoecology is germane here. All organisms are limited in their geographic distribution by basic ecological parameters which cannot be exceeded. Potential limiting parameters for any given organism are many--the basic task of the paleoecologist is to decipher which of these potential parameters were operative in a particular place during a particular span of time for the organisms under study. Correct identification of the parameters governing distribution of those organisms places definite constraints on what the nature of that environment could have been; the more parameters identified, the stronger the constraints and the more exactly defined the paleoenvironment.

As a highly simplified example, if subjection of an organism to temperatures greater than 30°C invariably resulted in death, the occurrence of that organism would imply that the microenvironmental temperatures where it lived never exceeded 30°C. We then would have an ecological parameter for that habitat. Another organism that could not survive freezing temperatures would reveal another such parameter; thus the maximum temperature fluctuations would be within the range 0-30°C. A third organism that required a particular food plant would not only add another aspect of the environment (the presence of that plant), but also give rise to the possibility of recognizing further parameters associated with that plant. Other organisms might add no data to this environmental reconstruction, however. One which is known to survive only within the temperature range of -10°C to 32°C would impart no further information here, since it could exist under the conditions demarcated by the other organisms of the example. The same organism, however, might be useful to set temperature parameters if it occurred at a different site.

With a large biota whose members' parameters were known, a quite exact and thorough delineation of the environment would be possible. The rub lies in recognizing the limiting parameters of organisms--such are known with any exactitude for few living species. And it must be the still-living taxa that are used as the primary source of data to set the environmental parameters, because the estimation of such parameters for extinct forms is, at best, indirect. Morphology (which necessarily is correlated to considerable degree with function) and association with taxa with "known" limiting factors are virtually the only methods for characterization of the needs of extinct taxa.

Theoretically important parameters that may be of significance include: 1) climatic, such as temperature extremes, seasonal distribution of precipitation, length of growing season, humidity, and the like; 2) geographic features, such as topographic relief and substratum type; 3) biological features, such as availablility of food, nesting sites, and cover, or the presence of competitors.

The limiting parameters are, of course, those to which the organism itself is actually subjected. Much of the success of the birds and mammals is due to their ability to avoid many of the gross environmental extremes, particularly by behavioral stratigems. High temperatures, for example, may

be avoided by taking shelter in burrows and limiting activity
to the crepuscular or nocturnal hours. Likewise, extremely
harsh winter conditions are avoided by many mammals by hiber-
nating in protected dens or by restriction of activity to
within the relatively warm mantle of snow.

The gross environmental conditions of a particular time
and place often are very different from those actually
endured by a particular organism in its microhabitat. This
is immaterial, however, if there is a direct linear correlation between the microenvironmental conditions to which an
organism is subjected and the gross environmental conditions.
To go back to the example of an organism that cannot survive
temperatures of greater than 30°C (determined by inspection
of gross environmental temperatures), it would be unimportant
for reconstructive purposes if the actual temperature the
organism succumbs to is really anything greater than 22°C as
long as a macroenvironmental temperature of greater than 30°C
results in the organism's habitat rising to a temperature
greater than 22°. The gross environment would still be
interpreted as greater than 30°C.

A somewhat different matter, and potentially more serious,
is the case where a parameter is compound. For example, a
dry heat of a certain temperature may be limiting whereas the
same temperature with greater humidity may be endurable.
Neither the humidity nor the temperature levels would be
stable parameters by themselves, the limiting feature being a
non-linear interplay between the two.

To a certain extent, dangers of these types may be alleviated by seeking out a wide variety of modern, varying conditions in attempting to determine the parameters of extant
taxa, but, as seen further on, some Pleistocene conditions
probably are not duplicated anywhere today. The solution, to

the extent that it exists, would seen to lie in consideration of the entire biota of a site along with such evidence as may be available from the fields of geology, climatology, geophysics, and the like. Only by viewing the evidence as a whole can the past environment be reconstructed into an internally consistent model. To use another simplified example, that of an organism existing in the past at higher temperatures than found within its range today because of the presence of higher humidity, the higher humidity should also have allowed organisms to coexist in that biota whereas they do not coexist today, as well as possibly showing up in the form of increased stalagmitic activity, possibly in preserved geomorphic features, or in the geochemical makeup of the deposits. Such integrated studies are in their infancy, and most of the sites treated in this work lack many of the data important for paleoecologic synthesis. Nevertheless, this present work is an effort to apply at least the philosophy to a large-scale reconstruction.

There are other drawbacks to good paleoecologic reconstruction: the nature of the sites themselves, the excavation and interpretation of the sites, and the identification of the recovered biota. Few sites follow the textbook examples of being deposited in even, clear, datable layers. Most of the more important sites dealt with here are cave deposits, which are notoriously complex as a class. Rockfalls, access routes opening and closing, material sifting or washing from level to level, and animals burrowing through the fill are just some of the factors complicating stratigraphy and interpretation. Moreover, cave deposits usually are not directly traceable to deposits outside the cave system, where they might be directly relatable to a better known geologic situation. The non-cave sites usually are much more restricted faunally and few are incapable of being misinterpreted. Although radiocarbon dating has helped chronologic interpretations immensely, there are too many potential sources of error to make it a panacea.

Excavation techniques have varied both through time and among excavators. In earlier days, some workers were interested only in the large, dramatic extinct megafauna, and smaller forms tended to be saved only when figuratively thrust upon them. Some workers have virtually ignored stratigraphy and geologic data in general, the animal specimens being the only things of importance. And even the most careful modern scholar may misinterpret relationships of chaotic deposits, particularly under oft-times trying excavation conditions.

The process of identification is not without its pitfalls. Usually working with fragmentary specimens and often with inadequate comparative materials, most paleontologists have made at least their share of blunders. Much of the published material has never been reexamined since the original publication. Moreover, in a study of this sort, utilizing results of studies spanning nearly a century, nomenclatural obscurities constantly obfuscate the taxa involved.

In summary, not only are there drawbacks in the application of paleoecologic reconstruction theory itself, but the data base to which it is applied has serious drawbacks. Only by attempting to construct the most internally consistant model possible with the methods and data available and then testing the model by applying refined methodology and new data can we hope to approach understanding. It is the first step, as inadequate as it may ultimately prove, that is attempted here.

III. PRESENT CONDITIONS

GEOGRAPHY AND CLIMATE

Any tendency to think of "The West" as an area unified by
climate and geography is eroded quickly by the barest contact
with reality. Our area covers some 3200 km north to south
and 1900 km east to west (Fig. 1). Elevationally, the area
extends from 86 m below sea level in Death Valley to about
4400 m in the Sierra Nevada and Rocky Mountains. Precipitation ranges from less than 50 mm in Death Valley to an annual
average of more than 3.8 m in the mountains of the Olympic
Peninsula of Washington. Other geographic and climatic
variables differ as widely. Four major river systems (the
Mississippi-Missouri, Rio Grande, Colorado, and Columbia)
drain the area, while much of the region has no drainage to
the sea. In brief, the region is extraordinarily diverse and
complex, providing examples of a wide variety of physical
conditions and the setting for an equally diverse biota.

As a basis for discussion, the area can be divided into
eight geologic provinces. Much of the following sketch is
derived from Shimer (1972). The Pacific Border Province
basically lies between the Pacific Ocean on the west and the
Sierra Nevada-Cascades on the east. It thus includes the
coastal ranges of Washington, Oregon, California, and the
mountains of Baja California as well as the great interior
valleys of the three former states. Glaciers still are

present in the Olympia Mountains of Oregon, which rise to almost 2440 m. Farther south, in the Klamath Mountains, "a few glacierets have a precarious existence at altitudes of 2500-2600 m" (Wahrhaftig and Birman, 1965).

Bounding the Pacific Border Province on the east are the Cascade and Sierra Nevada provinces. The former is an up-lifted area supporting a number of major volcanic cones from Canada into northern California; the recently noteworthy Mount St. Helens is just one of these. The Sierra Nevada, to the south, rises to 4418 m at Mt. Whitney, the highest peak in the conterminous USA. Some three score glaciers still occur in this range, and much of the higher elevation has been shaped extensively by former glaciation.

East of the Cascades, the Columbia Plateau covers south-eastern Washington, most of eastern Oregon and southern Idaho, and a sliver of northern Nevada. The region essen-tially is a plateau built from intermittent lava flows which, for the most part, pushed up quietly through fractures in the substatum. Soils mostly are of weathered volcanic material, but thick loess in the northeastern portion forms the Palouse section. A vast portion in the north, the Channeled Scab-lands, displays spectacular now-dry channels, waterfall plunges, and other flood features formed by catastrophic draining of huge ice-dammed lakes following failures of the dams.

The Basin and Range Province runs south from the Columbia Plateau into northern Mexico and across the southern halves of Arizona and New Mexico to the western edges of the Great Plains; most of Trans-Pecos Texas is included. The province is an area of generally north-south trending fault-block mountain ranges separated from one another by valleys. Ex-cept where the province is traversed by the Colorado River

and the Rio Grande, most of the valleys lack external drainage. Thus these basins consist largely of materials carried
into them from the surrounding highlands. Valley floors in
some cases (Death Valley, Imperial Valley) lie well below sea
level.

Many of the basins support playas, temporary lakes which
hold water only emphemerally. Many of the same basins show
clear evidence of permanent, deeper lakes during glacial
ages. A few permanent lakes persist today, among them the
Great Salt Lake, Pyramid and Winnemucca lakes, and Mono Lake.
In the Great Basin portion of the province, more than 150
basins occur; almost all held water during glacial/pluvial
times.

The Colorado Plateau Province covers most of northern
Arizona, the southeastern half of Utah, the northwestern part
of New Mexico, and a small section of southwestern Colorado.
This area is a high plateau, with most portions lying at over
1525 m elevation; portions extend to more than 3350 m. In
general, more or less horizontal strata characterize the region, with faults or narrow tilted areas between plateaus of
varying elevation.

The Grand Canyon forms a major barrier in the western part
of the province, being incised about 1.6 km into the plateau.
Volcanic action has been notable south of the canyon, around
Flagstaff and continuing south and east into New Mexico.
North of the canyon, sedimentary outcrops are more common,
though volcanism appears locally. The generally horizontal
strata and arid climate results in cliff-walled canyons sub-
dividing much of eastern Utah and adjacent Colorado. Most of
the province drains to the Colorado River, though the southeastern portion intersects the Rio Grande.

The Rocky Mountain Province extends from north-central New

Mexico generally northwesterly, bounding the Colorado
Plateau, the northeasternmost Basin and Range Province, the
Columbia Plateau, and the northeastern part of the Cascade
Province.

Shimer (1972) divided the province into four parts. The
Southern Rockies are centered in Colorado, with overlap into
northern New Mexico (San Juan-Jemez and Sangre de Cristo
mountains) and southern Wyoming (Laramie and Medicine Bow
ranges). The ranges trend north-south, with basins, such as
the San Luis Valley and North Park, between. A number of
peaks exceed 4250 m, with Mt. Evans reaching 4399; the intermontane basins generally lie over 2130 m. Numerous glacial
structures are apparent. Separation from the Great Plains to
the east is abrupt.

The Wyoming Basin subsection separates the Southern
Rockies from the Middle Rockies. This basin is an essentially non-mountainous area in northwestern Colorado and
central and southwestern Wyoming, connecting to the Great
Plains by the passageway between the Laramie Range to the
south and the Bighorn Mountains to the north.

The Middle Rockies include the Bighorn, Wind River, Uinta,
Beartooth, Absarokas, Grand Tetons, and Wasatch ranges. Many
of these ranges reach to elevations over 3650 m, while some
approach 4250 m. The Big Horn Basin is of notable size and
is similar in topography to the Wyoming Basin area.

The Northern Rockies continue northward into Canada from
the Middle Rockies. In central Idaho, a number of ranges
lack the linearity seen in most of the Rockies; these include
the Salmon River, Clearwater, Coeur d'Alene, and Sawtooth
groups. To the north of this area, linearity is again
notable, with the Rocky Mountain and Purcell trenches being
particularly prominent and extensive valleys. During the

Pleistocene, glacial lakes formed intermittently in these valleys as ice damming occurred. Small mountain glaciers still are active in some areas. The southeastern part of the Northern Rockies, in southwestern Montana, consists mostly of heavily eroded fault-block mountains.

The Great Plains Province borders the Rockies on the east, from well into Canada on the north to the eastern part of the Trans-Pecos on the south. The western boundary is high, generally around 1525 m, and the plains slope overall to the east. Although generally with little or moderate relief, some areas are marked by extensive incision, while others have outlying mountain masses surrounded by plains topography. The Canadian portion and the U.S. portion north of the Missouri River in Montana have been glaciated.

Montana shows a number of outlier ranges: Big Belt, Little Belt, Little Rockies, Big Snowy, Crazy, Bearpaw, and Judith mountains among them. Farther south, in southeastern Colorado and northeastern New Mexico, volcanic activity has resulted in considerable relief.

Climate varies tremendously within the study area. Major causes of climatic variability include the vast latitudinal and elevational spans, distance from moisture sources, orographic rain shadows, character of offshore oceanic currents, and the like. Climatic data are summarized in a number of publications, such as that by the National Oceanographic and Atmospheric Administration (NOAA, 1974). However, much of the West is sparsely populated by weather data-collecting stations, and many of the more extreme conditions normally go unrecorded. This is doubly the case in Mexico.

Climatic data frequently are misleading in implying that average figures are those of importance. It is believed, however, that extremes govern biological distributions in many cases. Thus for an organism utterly unable to withstand freezing temperatures, even one such event over a several-year span is sufficient to bar that organism from an otherwise suitable area.

Comparison of precipitation maps with physiographic maps shows dramatically the effect of mountain masses in increasing moisture amounts in the highlands; major ranges, such as the Sierra Nevada and Cascades, also drastically decrease precipitation to the leeward (the "rain shadow" effect). Not all aridity is due to such orographic effects, however; the extreme dryness of the Sonoran Desert is due in large part to a combination of a cold off-shore oceanic current and the latitudinal position under the sinking air masses of the horse latitudes.

Equally important as the overall amount of precipitation is its timing. Most of the region is characterized by seasonality of precipitation, with pronounced dry and wet seasons. Generally, summer precipitation is important along the eastern border of our area, although the peak precipitation time tends to migrate from July in the south to May or June farther north. With progression westward, however, emphasis switches away from the summer peak until an area of extreme summer drought is reached along the western boundaries.

Such seasonality of precipitation can be of great biological importance. The effect of a given amount of precipitation occurring during the summer-high temperatures of New Mexico is quite different from the same amount occurring during the cool months of western Nevada. To use one example, the yellow-bellied marmot is believed to be excluded from the lowlands (below about 3050 m under normal conditions) of New Mexico because of the lack of sufficient spring fodder to carry the animal over to the summer rainy period (Harris, 1970a); in south-central Nevada, however, Hall and Kelson (1959) recorded the animal as low as 2225 m.

Although precipitation is a critical feature in much of the West, other climatic features also hold great importance. For example, the mean length of the frost-free period is an approximate measure of the length of the growing season. This, of course, can be of prime importance in setting the character of the vegetation which, in turn, may be strongly selective of animal life. Within the western USA, the freeze-free period varies from over 330 days around Yuma, Arizona, to less than 30 days in a number of high mountain areas.

Mean average temperatures give an idea of the general temperatures of an area, but, in common with most averages, may be highly misleading. An average annual temperature of 10°C along the coast of California is a far different climatic habitat than is southeastern Colorado at the same mean temperature. In general, inland areas and areas of high elevation tend to be less equable (have greater extremes) than do coastal regions and those of low elevations.

MODERN VEGETATION

The modern vegetation of our area is extremely complex. The approach here is to simplify presentation somewhat by grouping the vegetational communities into four major groups that have some (though imperfect) internal functional consistency. The U.S. portions are shown in four separate diagrams adapted from Kuchler's (1970) map of potential natural vegetation. This is the vegetation that is believed would eventually result if man's influence were removed and the landscape allowed to revert to natural conditions.

The vegetational groups are forest (Fig. 2), woodland-shrubland (Fig. 3), shrubsteppe (Fig. 4), and grassland (Fig. 5). Within the major groups, some of Kuchler's vegetational types have been grouped together, hopefully in a compatible manner. This has been done to prevent maps that would be overly-cluttered for purposes here while still allowing presentation of major sub-groupings. One point should be stated strongly: these maps show communities dominated by the types of vegetation noted, but there is much intermingling of species and many taxa characteristic of one of these vegetative types also are widespread in other-dominated vegetational types (this is particularly true for grasses). Also, the West is notorious for local differences due to slope, substratum type, etc.

Occurrence of forest vegetation in the West (Fig. 2) shows a very high correlation with mountain systems. Orographic precipitation coupled with the cooling effect of increased elevation act to allow forest development where otherwise effective aridity would be too great. Thus the Rockies, the higher mountains of the Basin and Range Province, and the Sierra Nevada-Cascade areas are boldly delineated. Added to the mountainous regions are the wet coastal areas of the Pacific Northwest.

The lower-elevation inland forested sites generally are occupied by somewhat open forests of warmth- and arid-tolerant forms, particularly ponderosa pine. With increasing elevation, trees less and less tolerant of warm, xeric conditions appear until, at the upper tree-line, temperature and shortness of the growing season eliminate tree forms entirely. Thus the pine and mixed conifer forests are replaced by

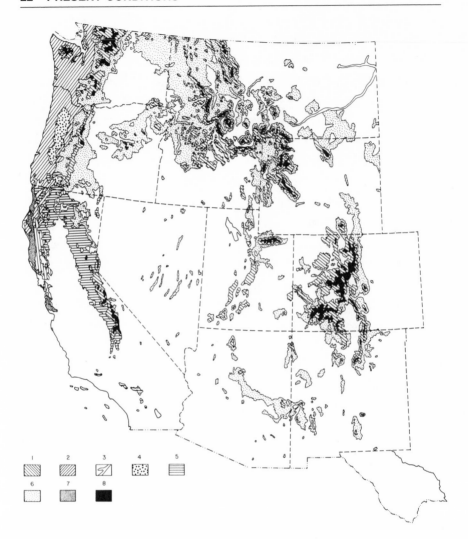

Fig. 2. Modern potential forest vegetation. Many of the demarcated units are of combined forest types. 1 = spruce-cedar-hemlock; cedar-hemlock-Douglas fir. 2 = fir-hemlock; western spruce-fir; spruce-fir-Douglas fir; southwestern spruce-fir. 3 = northern floodplain forest. 4 = Oregon oak-wood and mosaic of Oregon oakwood and cedar-hemlock-Douglas fir. 5 = mixed conifer; red fir. 6 = silver fir-Douglas fir; western ponderosa; Douglas fir; cedar-hemlock-pine; grand fir-Douglas fir; eastern ponderosa; Black Hills pine forest; pine-Douglas fir; Arizona pine forest. 7 = redwood. 8 = alpine meadows. Adapted from Kuchler (1970).

spruce-fir forests akin to the northern boreal forest that
spans the continent in central Canada, with alpine meadows
(alpine tundra) taking over at the highest elevations.

The mesic Pacific Northwest supports lush coniferous forest at virtually all elevations up to timberline. Included
are the spruce-cedar-hemlock and the cedar-hemlock-Douglas
fir forests at low to intermediate elevations and the silver
fir-Douglas fir and fir-hemlock forests of the higher elevations of Washington and Oregon. In northern and central
California, the magnificent coastal redwood forests occur.

In Fig. 3, the pattern of woodlands and shrublands is far
different, with a vast, nearly contiguous area centering in
the Great Basin and the lowlands of the Colorado River. In
some ways, this figure contains the most diverse of the vegetational types treated here. Nevertheless, in common is the
arid to semi-arid aspect with, usually, a notably seasonal
distribution of precipitation.

Much of the shrubland area qualifies, at least by popular
concept, as desert. Thus the Lower Colorado area includes
the Sonoran and Mojave deserts, dominated by creosotebush
(Larrea divaricata). To the north, the cold-desert Great
Basin is dominated by big sagebrush (Artemisia tridentata)
shrubland and, in the more alkaline areas, by greasewood
(Sarcobatus vermiculatus) and saltbush (Atriplex spp.).

The woodland component consists of low, generally widely-
spaced trees. These woodlands usually lie immediately below
forest in areas with enough relief to support such. In some
areas, particularly in the eastern half, topography likely
plays a large part in determining whether woodland or grassland dominates--areas of rough topography (with resultant
shallow, rocky soils) favor woodland growth. In the warm
south, the woodlands, or pygmy forest, are dominated by a

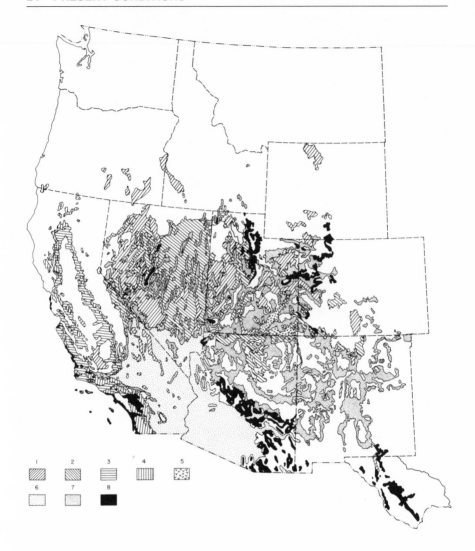

Fig. 3.   Modern potential shrubland and woodland vegetation. 1 = saltbush-greasewood. 2 = Great Basin sagebrush.
3 = Californian oakwood. 4 = chaparral. 5 = blackbrush.
6 = creosotebush; creosotebush-bursage; palo verde-cactus
shrub. 7 = pinyon-juniper woodland. 8 = (CA) coastal
sagebrush; (NV, UT, WY, CO) mountain mahogany-oak scrub; (AZ,
NM, TX) oak-juniper woodland; transitional between oak-
juniper and mountain mahogany-oak scrub. Adapted from
Kuchler (1970).

wide variety of live oaks (Quercus spp.). Farther north,
pinon-juniper woodland replaces the oak woodlands. Various
species of small pines (Pinus edulis, P. cembroides, P.
monophylla) and junipers (Juniperus pinchotii, J. monosperma,
J. utahensis) make up the dominant forms, the particular species varying geographically with temperature and precipitation patterns.

Figure 4 has two major subdivisions. In the south, the
major shrub component is creosotebush; unlike the creosote-
bush-dominant Sonoran and Mojave deserts, however, various
grasses often form a substantial proportion of the vegetation. The Chihuahuan Desert of southeastern Arizona, southern New Mexico, and Trans-Pecos Texas encompasses a substantial portion of this type. In the north, the dominant shrub
occurring with the steppe vegetation is sagebrush (Artemisia
spp.).

Figure 5 shows the distribution of grassland. Generally,
these areas receive more moisture (or more effectively-
distributed precipitation) than the shrub-steppes. The en-
croachment of the Great Plains into the eastern margin of our
area is clear.

In northern Mexico, we find essentially an extension of
the trends seen near the U.S. border (Brown and Lowe, 1980).
The Sonoran Desert continues south of the border and west of
the Sierra Madre Occidental. The latter supports vegetation
similar to that in the mountains of southwestern New Mexico
and southeastern Arizona: evergreen woodland at lower elevations is replaced by coniferous forest at higher elevations.
East of the Sierra Madre, grasslands and desert grasslands
give way to the desert-scrub of the Chihuahuan Desert, though
this is interrupted frequently by desert grasslands and occasionally chaparral and evergreen woodland on the desert
ranges rising from the lowlands.

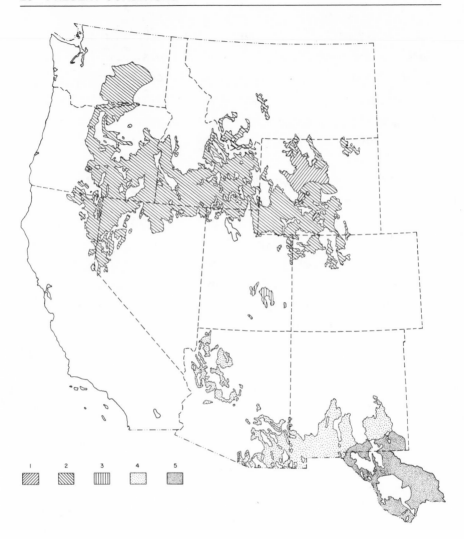

Fig. 4. Modern potential shrub and grassland combination vegetation. 1 = wheatgrass-needlegrass-sagebrush shrub-steppe. 2 = sagebrush steppe. 3 = galleta-three awn shrub-steppe. 4 = grama-tobosa-creosotebush shrub-steppe. 5 = Trans-Pecos tarbush-creosotebush shrub savannah. Adapted from Kuchler (1970).

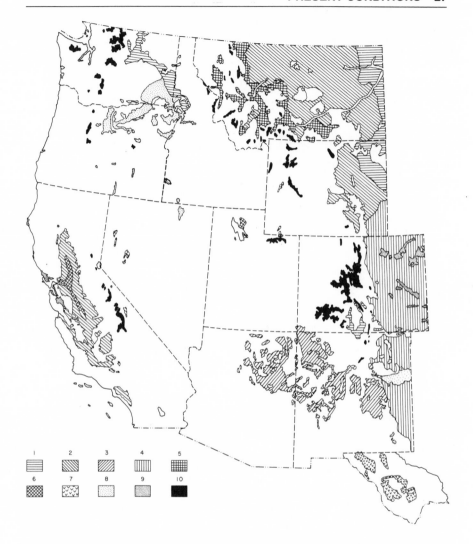

Fig. 5. Modern potential grassland vegetation. 1 = (WA, ID) fescue-wheatgrass; (MT, WY) wheatgrass-needlegrass. 2 = (MT, WY) grama-needlegrass-wheatgrass; (CA) Californian steppe (Stipa). 3 = grama-galleta steppe. 4 = grama-buffalo grass. 5 = foothills prairie. 6 = tule marshes. 7 = grama-tobosa prairie. 8 = (WA, OR, NV) fescue-wheatgrass; (NM) oak-bluestem shinnery. 9 = (WA, OR, ID) wheatgrass-bluegrass; (CO) sandsage-bluestem prairie. 10 = alpine meadows. Adapted from Kuchler (1970).

The Canadian vegetation likewise can be viewed as predominantly an extension of the vegetation south of the U.S.-Canadian border. Prairie vegetation extends northward far into the "Prairie Provinces" of Alberta and Saskatchewan, eventually grading into aspen savannah and then into coniferous forest. In similar fashion, the high-mountain forests of the Rockies continue into the highlands of Canada and the Northwest Coast forestlands continue up the west coast of Canada.

IV. THE PLEISTOCENE IN GENERAL

CHRONOLOGY

Ages of Late Pleistocene events, particularly those beyond
the range of radiocarbon dating, have not received universal
agreement. For example, Kurten and Anderson (1980) suggested
a date of about 300,000 years BP for the beginning of the
Wisconsinan. However, Broecker and van Donk (1970) indicated
that sea level maxima occurred at around 82,000, 103,000, and
124,000 BP; these likely represented events of the last
interglacial, the Sangamonian. Broecker and van Donk also
noted evidence of major glaciation between 50,000 and 80,000
BP (presumably early Wisconsinan glaciation) and that at
least at times during the span between this glaciation and
the Wisconsinan full-glacial peak at about 18,000 BP, ice
sheets were present in the northern hemisphere and larger
than half of their maximum size.

   Although presence of oscillations between more glacial and
less glacial (stadial, pluvial; interstadial, interpluvial)
conditions within the Wisconsinan obviously is of biological
interest, there is little solid information surely relating
the more detailed marine record (Broecker and van Donk, 1970;
Emiliani and Shackleton, 1974) to specific terrestrial
faunas. Those faunas here labeled as interstadial are
recognized by their departure in character from full-glacial
faunas. However, not only is assignment to a particular

interstadial interval difficult or impossible considering the
reliability of the dating available, but chronologic placement of a fauna within such a span presently is impossible.

Recognized interstadial faunas from our area are rare.
Wendorf (1975) noted an interpluvial (the Rich Lake Interpluvial) in the Llano Estacado, just east of our area, with a
date at 26,500 ± 800 BP, presumably correlated with an interpluvial at Clear Creek, Texas, dated at 28,840 ± 4730 BP.
These dates jibe reasonably well with dates of the three Dry
Cave interstadial sites (radiocarbon dated on notoriously
untrustworthy bone carbonates) of 25,160 ± 1730, 29,290 ±
1060, and 33,590 ± 1550 BP, though one or more of these may
be correlative with an earlier interpluvial (Arch Lake Interpluvial) (Wendorf, 1975). Even assuming these correlations
are correct, the Dry Cave faunas could range from anywhere
within the interpluvial cycle as conditions changed from near
pluvial to most extreme interpluvial and back to near pluvial
again. Even worse, from the viewpoint of clear interpretation, the stades of the "mid-Wisconsinan," lying between the
major early Wisconsinan glaciation and the late Wisconsinan
full-glacial, apparently were of much lesser severity; in the
absence of a large sample of mid-Wisconsinan faunas, we can
distinguish only between full-glacial faunas and faunas that
are not of full-glacial character--the "interstadial" faunas
of this study in actuality could represent any climatic mode
within the mid-Wisconsinan from most to least glacial. Thus
it must be kept in mind that although the term interstadial
is used here, it refers only to one or more unknown phases
within the mid-Wisconsinan climatic fluctuations.

The Wisconsinan full glacial is known to be complex in
itself, at least during the waning stages. A major peak of
the late Wisconsinan stade apparently occurred around 12,000

BP (Broecker and van Donk, 1970; Wendorf, 1975), after less-
ening of the peak stadial (full-glacial) climatic conditions
that had followed the mid-Wisconsinan. This was followed by
rapid amelioration to around 11,000 BP. Evidence from a number of areas (Van Devender and Spaulding, 1979; Broecker and
Kaufman, 1965) indicates modern vegetative and climatic conditions, however, did not appear until close to 8000 BP or
later. In much of the West, a warm Altithermal interval followed shortly thereafter, lasting 2000 to 3000 years and being replaced by approximately modern climatic conditions.

VEGETATION

Wisconsinan vegetation reconstruction primarily rests on
palynologic (pollen and spore) analysis and on occurrences of
macrofossils. In recent years, datable macrofossils from
woodrat (Neotoma) middens have been used extensively.

Palynologic study relies on the recognizability to a useful taxonomic level of the highly resistant exocrine layer of
pollen and spores. Such pollen grains commonly are preserved
in lake and bog sediments and, to a generally lesser degree,
in drier depositional environments. Drawbacks to interpretation include: 1) dependence almost entirely on anemophilus
(wind-distributed) pollens; 2) production of different
amounts of pollen by different kinds of plants and by plants
of the same kind under different environmental conditions; 3)
different ease of dispersal by wind (that is, some kinds are
deposited quite close to the source plants while other kinds
are carried enormous distances); 4) differential preservabil-
ity; 5) limitations as to level of identification, many kinds
being identifiable only to suprageneric groups, generic

groups, or species groups; 6) local concentration by natural forces (e.g., floating pollen being blown to one side of a pond before sinking to the bottom); 7) overrepresentation of some types because of the environment of deposition (e.g., wet-land types from plants growing in the immediate area of deposition); 8) usually the necessity for use of only relative frequencies per depositional level, rather than absolute frequencies per time unit (thus a relatively few pollen grains of pine transported from highland forests by wind into a desert situation may nevertheless form an appreciable portion of the deposited pollen due to poor pollen production within the desert; in a case of increasing aridity causing decreasing local pollen production while pine pollen influx remains steady, the pollen analysis would show an increasing percentage of pine--the increase in pine percentage could well be interpreted as showing increased moisture).

Palynologists generally are well aware of these drawbacks and attempt to correct for them as best as possible. Usually, pollen is considered as presenting evidence of broad regional vegetations with considerable accuracy, but as lacking the resolution to present finely-detailed pictures. Advances in theory and technique (e.g., Jacobson and Bradshaw, 1981; Leopold et al., 1982), however, hold hope for better separation of regional and local data.

Macrofossils in general depositional environments have the advantage of often being identifiable to low taxonomic levels and of demonstrably having occurred within a more circumscribed area than is the case with pollen. However, wood, leaves, and fruiting structures may be carried for long distances by running water. Where a past drainage basin was large, macrofossils may have been carried in from distant highlands or from upstream areas hundreds of kilometers distant from the site.

Woodrat middens have the advantage of normally containing plant materials from within only a short distance (ca. 100 m; Van Devender and Everitt, 1977) of the site. Thus, aside from possible rare pickups of transported material, the rodent-collected material gives a picture of strictly local vegetation that, moreover, is easily radiocarbon dated. The major disadvantage is that these middens are preserved almost exclusively within rock shelters and caves, leaving without representation those regions and local habitats that lack such sites.

Until relatively recent years, there has been a strong tendency for authors, particularly those whose primary focus has been on vertebrate paleoecology, to assume that Pleistocene pluvial vegetation acted in a simplistic manner to the simplistic climatic model of decreased temperatures and increased effective moisture: i.e., that northern plant communities advanced southward; high-elevation plants descended enmass to lower altitudes; and plants now characteristic of warm, arid regions retreated southward and to lower ground. Thus Western life zones were displaced, bodily as it were, little or no different from those of today if the presence of extinct animals was disregarded.

This tendency lingers on in later work in paleoecology despite the alarum bells that warned of greater complexity. These include the association in Kansas and elsewhere (Hibbard et al., 1965) of animals now far allopatric, the pointing out that modern marginal populations exist in non-typical habitats (Harris, 1977b), and the increasing information from pollen studies that plant communities during the late Wisconsinan often were of composition unmatched anywhere today (e.g., Cushing, 1965).

Within the past few years, particularly with the addition of information from woodrat middens, it has become very clear that the vegetational history is extremely complex. What follows is a brief run-down from the literature of major patterns seen in the West; evidence from vertebrate faunas will be considered later.

In the Pacific Northwest, Heusser (1965) interpreted palynological data (from a limited number of sites) as indicating the presence of lodgepole pine parkland during the late Wisconsinan. Lesser numbers of other conifers (including western white pine in the Puget lowland; Sitka spruce, true fir, and western white pine in the northern Willamette Valley of Oregon; and ponderosa pine in east-central Washington, accompanied by a considerable amount of grass) occurred with the lodgepole pine.

Poorly dated non-glacial (pre-late Wisconsinan maximum) deposits from the Puget lowland indicate "within the limits of pollen recognition, vegetation components during the interglacials did not differ from those growing at present in western Washington" (Heusser, 1965:479).

In the Southern Rockies area, much of the palynological work sheds little light on floral composition, that being considered as having been similar to present communities (though at different elevations). However, estimates have been made of treeline depression. Weber (1965:457), citing Maher (1961, 1963), suggested that in the San Juan Mountains of southwestern Colorado, "tree-line stood at least 600 m lower than now" at some time before 13,500 BP.

Modern distribution suggests the Cordilleran flora extended to lower elevations and eastward while "the entire eastern

woodland flora extended westward, probably along the major
watercourses, and mingled with the Cordilleran flora along a
wide area of the western Great Plains" (Weber, 1965:457).
Relicts of the latter occur now along the eastern mountain
front and of the former in the Black Hills and elsewhere in
the Great Plains. Phytogeography also suggests past connections to the northwest, since broken by the arid reaches of
southern Wyoming (Weber, 1965).

Martin and Mehringer (1965) published a diagram of hypothetical full-glacial vegetation from southern California
east to western Texas. In broad pattern, they suggested expansion of life zones now restricted to higher elevations
into the lowlands, with ponderosa pine parkland extending
into what today is upper desert and with pinyon-juniper woodland covering most of what now are the middle-desert elevations. Desert was limited to what currently is lowest Sonoran Desert. Slightly higher, but below woodland, was pictured
sagebrush or chaparral growth.

Woodrat midden studies published in a number of papers by
Wells et al. and by Van Devender et al. reveal more detail.
Desert apparently was even more restricted (and perhaps absent from the U.S. proper), with juniper woodland extending
to extremely low elevations. Betancourt and Van Devender
(1981) noted also the absence of ponderosa pine at all elevations studied from the area east of the California highlands,
implying that present populations to the east are late inva-
ders. They hypothesized that most of the area from somewhere
within the current pinyon-juniper zone into current low desert was inhabited by either pinyon-juniper woodland or by a
more xeric juniper woodland. Since that paper, Van Devender
and Toolin (1983) have recorded ponderosa pine in one area of
the Southwest, at Rhodes Canyon in the San Andres Mountains

of south-central New Mexico. Wells, Van Devender, and co-
workers saw in most areas a telescoping together of current
woodland and sub-woodland plants, though with a general absence of typical low desert forms. In the San Juan Basin of
New Mexico, Betancourt and Van Devender (1981:657) found evidence at 10,600 and 9460 BP "of communities dominated by
Douglas fir, Rocky Mountain juniper (Juniperus scopulorum),
and limber pine (Pinus cf. flexilis)." The midden sites are
lower than present day local stands of pinyon "and on slopes
presently occupied by Great Basin desert scrub."

In the Frenchman Flat area of the southern Great Basin,
Wells and Jorgensen (1964) found abundant juniper (Juniperus
osteosperma) in woodrat middens located in low desert ranges
that now do not support woodland at any elevation. The middens (at elevations of 1100 to 1830 m) are in areas today
occupied by creosotebush and blackbush desert shrub communities. The present lower limit of woodland in the region is
about 1700 m, and it occurs only in the higher ranges. Pinyon (Pinus monophylla) was absent below 1550 m, leading Wells
and Jorgensen (1964) to characterize the lower woodland climate as "more arid than the usual pinyon-juniper climate."

In Trans-Pecos Texas and southern New Mexico, the lowest
areas apparently were occupied by woodland. Wells (1966)
found both pinyon (P. cembroides remota) and juniper (J. pinchotii) at the lowest midden available (600 m, at Maravillas
Canyon at the eastern end of the Big Bend). Van Devender et
al. (1977), Van Devender and Everitt (1977), and Van Devender
and Riskind (1979) have documented woodlands at low elevations in far western Trans-Pecos Texas.

Van Devender (1977) and Van Devender and Spaulding (1979)
have summarized much of the woodland evidence, not only for
the Chihuahuan Desert, but also for the Sonoran and Mojave

deserts. In the latter paper, descent of montane forests also was documented from the literature for such places as the
Guadalupe Mountains of Texas (spruce and Douglas fir at 2000
m), southeastern Utah (spruce and Douglas fir at 1710 m),
Clark Mountain in San Bernardino Co., California (limber pine
and white fir at 1910 m), and the Sheep Range in southern
Nevada (subalpine bristlecone pine-limber pine at 1900 to
2400 m). At higher elevations in New Mexico, spruce-fir forest extended to ca. 2000 m on the San Agustin Plains (Foreman
et al., 1959). Smartt and Harris (1979) found evidence of
spruce at 1555 m in extreme south-central New Mexico.

Wells (1979:315) noted not only that juniper was identified from as low as 258 m in the Sonoran Desert of California (Chemehuevi Mountains), but that juniper woodlands were
present "in the southern lowlands of the Sonoran Desert . . .
in subtropical Sonora, Mexico."

BIRDS AND MAMMALS OF THE PLEISTOCENE

This section is designed to allow the reader to put some type
of common name to the major groups used here.

Table 1 lists 3 orders of birds and 11 orders of mammals.
Table 2 lists the families (38) of birds and mammals considered and allows placement of each into its proper taxonomic
order. Table 3 lists the 108 genera represented and their
placement to family.

In these tables, there is a major departure from usual
scientific practice. Normally, such lists are ordered by
taxonomic relationships. However, locating unfamiliar taxa
in such a list is time consuming, and the reader seems better
served by an alphabetical listing; this has been done here.

TABLE 1. Taxonomic orders of birds and mammals considered.

---

 1. Artiodactyla. Even-toed Hooved Mammals.
 2. Carnivora. Carnivores.
 3. Chiroptera. Bats.
 4. Cuculiformes. Cuckoos and Roadrunners.
 5. Edentata. Armadillos and Sloths.
 6. Galliformes. Gallinaceous Birds.
 7. Insectivora. Shrews and Moles.
 8. Lagomorpha. Rabbits, Hares, and Pikas.
 9. Marsupialia. Opossums.
10. Perissodactyla. Tapirs and Horses.
11. Primates. Man.
12. Proboscidea. Mastodons and Mammoths.
13. Rodentia. Rodents.
14. Strigiformes. Owls.

TABLE 2. Families of birds and mammals considered.
Numbers in parentheses refer to the taxonomic orders in
Table 1. Families that are extinct in North America are
marked by †.

---

 1. Antilocapridae (1). Pronghorns.
 2. Aplodontidae (13). Sewellels.
 3. Bovidae (1). Sheep, Mountain Goats, Bison, and
        Oxen.
 4. †Camelidae (1). Camels.
 5. Canidae (2). Dogs, Wolves, and Foxes.
 6. Castoridae (13). Beavers.
 7. Cervidae (1). Deer.
 8. Cricetidae (13). Native Mice and Rats.
 9. Cuculidae (4). Cuckoos and Roadrunners.
10. Dasypodidae (5). Armadillos.
11. Didelphidae (9). Opossums.
12 †Elephantidae (12). Mammoths.
13. †Equidae (10). Horses.
14. Erethizontidae (13). Porcupines.
15. Felidae (2). Cats.
16. Geomyidae (13). Pocket Gophers.
17. Heteromyidae (13). Pocket Mice and Kangaroo Rats.
18. Hominidae (11). Man.
19. Leporidae (8). Rabbits and Hares.
20. †Mammutidae (12). Mastodons.
21. †Megalonychidae (5). Megalonychid Ground Sloths.
22. †Megatheriidae (5). Megathere Ground Sloths.

23. Molossidae (3). Free-tailed Bats.
24. Mustelidae (2). Skunks and Weasels.
25. †Mylodontidae (5). Mylodont Ground Sloths.
26. Ochotonidae (8). Pikas.
27. Phasianidae (6). Quail.
28. Phyllostomatidae (3). Leaf-nosed and Vampire Bats.
29. Procyonidae (2). Raccoons and Ringtails.
30. Sciuridae (13). Squirrels.
31. Soricidae (7). Shrews.
32. Strigidae (14). Typical Owls.
33. Talpidae (7). Moles.
34. Tapiridae (10). Tapirs.
35. Tayassuidae (1). Peccaries.
36. Tetraonidae (6). Grouse.
37. Ursidae (2). Bears.
38. Vespertilionidae (3). Vespertilionid Bats.

TABLE 3. Genera of mammals and birds considered. Numbers in parentheses refer to the families in Table 2. Extinct genera are marked by †; genera extinct in North America but extant elsewhere by *.

Aegolius (32). Boreal Owl, Sawhet Owl.
*Acinonyx (15). Cheetas.
Alces (7). Moose.
Ammospermophilus (30). Antelope Ground Squirrels.
Antilocapra (1). Pronghorns.
Antrozous (38). Pallid Bats.
Aplodontia (2). Sewellels.
†Arctodus (37). Short-faced Bears.
Baiomys (8). Pygmy Mice.
Bassariscus (29). Ringtails.
Bison (3). Bison.
Brachylagus (19). Pygmy Rabbits.
†Camelops (4). Extinct Camels.
Canis (5). Dogs, Wolves, and Coyotes.
†Capromeryx (1). Extinct Pronghorns.
Castor (6). Beavers.
Centrocercus (36). Sage Grouse.
Cervus (7). Wapiti.
Clethrionomys (8). Red-backed Voles.
Coendu (14). Porcupines.
Conepatus (24). Hog-nosed Skunks.
Cryptotis (31). Least Shrews.
Cuon (5). Dholes.

Cynomys (30). Prairie Dogs.
Dasypus (10). Armadillos.
Desmodus (28). Vampire Bats.
Dicrostonyx (8). Collared Lemmings.
Didelphis (11). Opossums.
Dipodomys (17). Kangaroo Rats.
Eptesicus (38). Big Brown Bats.
*Equus (13). Horses.
Erethizon (14). Porcupines.
†Euceratherium (3). Extinct Brush Oxen.
Felis (15). Cats.
Geococcyx (9). Roadrunners.
Geomys (16). Pocket Gophers.
Glaucomys (30). Flying Squirrels.
†Glossotherium (25). Extinct Mylodont Sloths.
Gulo (24). Wolverines.
†Hemiauchenia (4). Extinct Llamas.
Homo (18). Man.
†Homotherium (15). Scimitar Cat.
Lagurus (8). Sagebrush Voles.
Lasionycteris (38). Silver-haired Bats.
Lasiurus (38). Lasiurine Bats.
Leptonycteris (28). Long-nosed Bats.
Lepus (19). Jack Rabbits.
Liomys (17). Spiny Pocket Mice.
Lutra (24). River Otters.
Lynx (15). Bobcats and Lynxes.
Macrotus (28). Leaf-nosed Bats.
†Mammut (20). Mastodons.
†Mammuthus (12). Mammoths.
Marmota (30). Marmots.
Martes (24). Martens.
†Megalonyx (21). Megalonychid Ground Sloths.
Mephitis (24). Striped Skunks.
Microdipodops (17). Kangaroo Mice.
Microtus (8). Voles.
Mustela (24). Weasels.
Myotis (38). Mouse-eared Bats.
†Navahoceros (7). Extinct Mountain Deer.
Neotoma (8). Woodrats.
†Nothrotheriops (22). Shasta Ground Sloths.
Notiosorex (31). Desert Shrews.
Ochotona (26). Pikas.
Odocoileus (7). Deer.
Ondatra (8). Muskrats.
Onychomys (8). Grasshopper Mice.
Oreamnos (3). Mountain Goats.
Oreortyx (27). Mountain Quail.

Ovibos (3). Muskoxen.
Ovis (3). Sheep.
†Palaeolama (4). Extinct Llamas.
Panthera (15). Big Cats.
Pappogeomys (16). Pocket Gophers.
Perognathus (17). Pocket Mice.
Peromyscus (8). White-footed Mice.
Phenacomys (8). Heather Voles.
Pitymys (8). Pitymyine Voles.
†Platygonus (35). Extinct Peccaries.
Plecotus (38). Big-eared Bats.
Procyon (29). Raccoons.
Rangifer (7). Caribou.
Reithrodontomys (8). Harvest Mice.
Scapanus (33). Moles.
Sciurus (30). Tree Squirrels.
Sigmodon (8). Cotton Rats.
†Smilodon (15). Sabertooth Cats.
Sorex (31). Long-tailed Shrews.
Spermophilus (30). Ground Squirrels.
Spilogale (24). Spotted Skunks.
†Stockoceros (1). Extinct Pronghorns.
Sylvilagus (19). Cottontail Rabbits.
†Symbos (3). Woodland Muskoxen.
Synaptomys (8). Bog Lemmings.
Tadarida (23). Free-tailed Bats.
Tamias (30). Chipmunks.
Tamiasciurus (30). Red Squirrels.
Tapirus (34). Tapirs.
Taxidea (24). Badgers.
Tayassu (35). Peccaries.
†Tetrameryx (1). Extinct Pronghorns.
Thomomys (16). Pocket Gophers.
†Tremarctos (37). Bears.
Urocyon (5). Gray Foxes.
Ursus (37). Bears.
Vulpes (5). Red Foxes.

In treating species, a different stratigem has been employed. The basic problem is that some species appear in
more than one context: 1) as modern forms used to elucidate
ecological parameters of Pleistocene sites, 2) as extant
forms deserving of comment regarding their Pleistocene occur-

rences, 3) as fossil forms whose ecology is being interpreted, and 4) as fossils whose occurrences need to be noted but which have not been utilized in interpretation nor require comment. Rather than scatter such accounts, with inevitable, space-wasting duplication (and frustration to a reader looking for a specific taxon), the first three categories (those requiring individual comments for each taxon) have been combined into Appendix 1. A complete list of bird and mammal taxa treated is presented in Appendix 2, with sites of occurrences coded for each taxon. Appendix 3 is a list of sites, including both those directly utilized and others which helped shape my picture of conditions, together with lists of taxa identified from each. In the appendices, as in the tables, presentation of the taxa is alphabetical.

Scientific names of extant mammalian taxa follow Jones et al. (1982) with the following exceptions: Brachylagus is used instead of Sylvilagus for the pygmy rabbit, Lynx instead of Felis for the lynxes, and Pitymys as the generic name for voles often recognized as subgenera (Pitymys and Pedomys) of Microtus (Repenning, 1983). Common names follow the same source and are given at the first occurrence in the text. Names of extinct forms are after Kurten and Anderson (1980) except in a few cases of taxonomic disagreement noted later. Extinct species and subspecies are noted throughout by †.

V. INTERPRETATIONS

SITE STANDARDIZATION

To interpret the paleoecology, some approach was necessary to make faunas from different sites comparable. There are two major problems: sites are scattered in time and they are scattered in space.

To solve the first problem for the late Wisconsinan interpretation, sites believed representative of time since the last mid-Wisconsinan interstadial were chosen for interpretation. Thus sites believed to date older than about 24,000 years were not considered. For many, ^{14}C dates were useful; for some, faunal makeup was utilized as a guide. A major problem which must be borne in mind is that several of the more important sites also included Holocene material and some (particularly Smith Creek Cave) may have included older material. Rejection of all but the "pure" sites would have made the sample too small for useful results, but the inclusion of contaminated data necessarily blurs the results.

Once a selection of sites was available, some method was necessary to allow for the effects of latitude, longitude, and elevation. Because the West Coast of North America tends to differ dramatically from areas east of the Sierra Nevada/ Cascade ranges, the West Coast interpretation is kept separate from that of the Interior. The latter (Interior Division) will be considered first.

Within the interior, there still is considerable east-west variation in temperature and, particularly, precipitation. This variablility has been ignored in the general treatment. The north-south temperature variation and the effect of elevation on temperature, however, are too severe to be ignored. To render these variations less critical, an artificial--but useful--convention is used.

Temperature tends to decrease with both increasing latitude and increasing elevation. Approximately the same effective decrease in temperature can be obtained by going northward 1° of latitude as by increasing elevation by 107 m. A locality at any latitude can be converted to a hypothetical elevation above sea level at the equator (keeping in mind that this is a highly artificial construct) by multiplying the latitude of the locality by 107 and then adding the site elevation in meters. If, for example, a site is at an elevation of 2000 m and a latitude of 40°, then its "corrected" elevation is 40 times 107 plus 2000, or the equivalent of 6280 m above sea level at the equator (equator equivalent, or ee).

This stratigem allows sites to be placed in rough relative positions to each other in terms of temperature. In real life, of course, local climatic conditions, topography, and other features insure that any such relationship is only approximate. In addition, sea level is known to have fluctuated significantly during the Pleistocene due to varying amounts of water locked within glacial ice. Thus the base line for calculating elevation ee has changed through time, but lack of fine chronologic control and uncertainties regarding the magnitude of sea-level depression at any given time prevent meaningful corrections for this factor at present.

FULL GLACIAL--INTERIOR DIVISION

Examination of the faunal and floral evidence from the Interior Division sites indicates several vegetative types occurred in a logical elevational progression (Table 4), though overlapping broadly. These were tundra, boreal forest, sagebrush, woodland, and grassland. Not only did these types appear in logical sequence, but they also fell into four or five major (though somewhat intergrading) zones.

Tundra is recognized here primarily by the presence of Dicrostonyx (lemmings) or Ovibos (muskoxen). Occurrence of Rangifer (caribou) or Oreamnos americanus (mountain goat) suggests tundra and, in the absence of the first two taxa, are used for a queried occurrence. Other species occurring today in tundra situations are not considered diagnostic since they also occur regularly at lower elevations. Tundra as used for the Pleistocene sites may be strictly equivalent with neither Arctic nor alpine tundra, but does imply a severity of climate such as to have placed the area above (or north) of the local treeline.

The boreal habitat is envisioned as having been similar in gross form to the combined Hudsonian and Canadian life zones of Merriam, the vegetation type popularly referred to as spruce-fir forest. Taxa used more or less as marker species include Ochotona princeps (pika) (which may be a tundra inhabitant at some sites), Gulo gulo (wolverine), Tamiasciurus hudsonicus (red squirrel), Phenacomys intermedius (heather vole), Tamias umbrinus (Uintah chipmunk), Sorex palustris (water shrew), Lynx canadensis (Canadian lynx), Rangifer (also a tundra inhabitant), Lepus americanus (snowshoe hare), Sorex hoyi (pygmy shrew), Clethrionomys (red-backed voles), and Mustela erminea (ermine). Numerous other species occur

TABLE 4. Full-glacial Interior Division sites used for interpretation, showing vegetative zone interpretation, elevations, and elevations ee for each site. Vegetation indicated by the extant fauna and by vegetation from middens is coded as 1 = tundra, 2 = boreal, 3 = sagebrush, 4 = grassland, 5 = woodland. Elevations in meters.

Site	Veg.	Elev.	Elev. ee
NORTHERN HIGHLAND ZONE			
Jaguar Cave	1,2,3,4	2255	7017
Horned Owl Cave	?,2,3,4	2439	6880
Bell Cave	1,2,3,4	2379	6820
Chimney Rock Animal Trap	2 4	2410	6797
Watino	1	ca. 600	ca. 6540
Cochrane	?	ca. 1067	ca. 6524
Granite Canyon No. 1	2	2070	6350
Little Box Elder Cave	1,2,3,4	1676	6224
Prospects Shelter	1,2,3,4	ca. 1400	ca. 6215
Natural Trap Cave	1,2,3,4	1400	6215
Little Canyon Creek Cave	1,2,3	ca. 1430	ca. 6138
Smith Creek Cave	2,3,?	1950	6123
Smith Creek Cave No. 4, No. 5	2,3	1950	6123
Empress	?,?	ca. 640	ca. 6097
Hell Gap Site	4	1525	6073
Douglas, WY	1	1468	6069
MIDDLE-ELEVATION SAVANNAH ZONE			
Streamview No. 1, No. 2	2,3	1860	6033
Sheep Range, South Crest Midden	2,3	1990	5896
Wilson Butte Cave	2,3,4	ca. 1311	ca. 5859
Agate Basin Site	?,3,?	1190	5845
Garrison No. 1, No. 2	2,3	1640	5813
Atlatl Cave	2,3	1910	5762
Sheep Range, Flaherty Mesa No. 1	2	1770	5676
Dutton	4	1255	5482
Isleta Cave No. 1, No. 2	3,4	1716	5461
Selby	4	1173	5453
Upper Sloth Cave	2 4,5	2000	5425
Lower Sloth Cave	2 4,5	ca. 2000	ca. 5425
Dust Cave	2 4,5	2000	5425
Mescal Cave	2 4	1550	5349
Howell's Ridge Cave	3,4,?	1675	5099

Site	Veg.	Elev.	Elev. ee
Muskox Cave	2, 4,5	1600	5024
Hermit's Cave	? ?	ca. 1600	ca. 5024
SAGEBRUSH STEPPE-WOODLAND ZONE			
Blackwater Draw Sites	4,5	ca. 1280	ca. 4972
Papago Springs Cave	?,5	1586	4957
Williams Cave	5	1495	4919
Burnet Cave	4,5	ca. 1435	ca. 4913
Shelter Cave	3 5	1435	4859
Hueco Tanks No. 1	4	1420	4844
Conkling Cavern	3,4,5	1399	4823
Stanton Cave	3	927	4779
Dry Cave Sites	3,4,?	1280	4758
Dark Canyon Cave	3,4,5	1100	4578
Tule Springs E-1	3	703	4555
STEPPE-WOODLAND ZONE			
Vulture Cave	5	645	4497
Gypsum Cave	?,?	610	4462
Schuiling Cave	5	658	4403
Rampart Cave	5	525	4377
? STEPPE ZONE			
Jimenez Cave	4	1450	4339
Ventana Cave	4	750	4228

today within this zone, but are not considered to be marker species. In this zone, as in all others, a combination of species non-diagnostic individually may be taken as indicative of the the boreal vegetative type having been present.

At one stage, an open coniferous forest type analogous to the ponderosa pine and ponderosa pine-Douglas fir open forests of the Southwest was considered. However, it is apparent that such a zone is recognizable faunalistically today primarily by the presence of _Sciurus aberti_ (Abert's squirrel) and _Microtus mexicanus_ (Mexican vole). Both of these now are associated closely with ponderosa pine--but this pine

appears to have been rare in the Wisconsinan Southwest.
Sciurus aberti is absent from the fossil record and M. mexicanus is known only from areas presumed to have lacked ponderosa pine and so must have had a different association during the late Wisconsinan (presumably today M. mexicanus is
dependent on grasses, climate, and cover currently associated
with ponderosa pine rather than with the pine itself). A
forest structurally similar to the present open coniferous
forests may well have been present, but faunalistically would
have merged into adjacent zones.

A sagebrush habitat was recognized on the basis of forms
now closely associated with Artemisia tridentata: Brachylagus idahoensis (pygmy rabbit), Lagurus curtatus (sagebrush
vole), and Centrocercus urophasianus (sagebrush grouse).

Recognition of woodland was mostly subjective on the basis
of joint occurrence of taxa tending to reach their higher
limits about this zone and taxa generally reaching their lower limits there. Peromyscus truei (pinyon mouse) is considered a marker species; Urocyon cinereoargenteus (gray fox) is
used similarly, but with considerable trepidation. Woodrat
midden data are of much aid in delineating the zone.

A grassland component was notable throughout the elevational range. Marker taxa used to recognize its presence
include Cryptotis (least shrews), Spermophilus richarsoni
(Richardson's ground squirrel), Geomys bursarius (plains
pocket gopher), Cynomys ludovicianus (black-tailed prairie
dog), Pitymys ochrogaster (prairie vole), and Vulpes velox
(swift fox).

The vegetative types fall into at least four and possibly
five recognizable complexes. From highest elevation to lowest, these are: 1) Northern Highland Zone, 2) Middle-
elevation Savannah, 3) Sagebrush Steppe-woodland, 4) Steppe-

woodland, and 5) Steppe. These, together with the sites
assigned to each, are discussed below.

Northern Highland Zone

The uppermost zone included all sites above about 6069 m ee
(Table 4). Most sites within this zone displayed evidence of
tundra, boreal forest, sagebrush, and grassland. The limits
of the zonal span are defined by the highest and the lowest
sites showing all four vegetational types. With one exception, those sites not suggesting tundra but included within
this elevational span either would have been isolated from
the ranges of tundra mammals by notably lower-lying terrain
(Granite Canyon No. 1 and the Smith Creek Cave sites) or are
in areas of relatively slight topographic relief (Cochrane,
Hell Gap Site, Watino). The single exception, easily ex-
plainable by sampling error, is Chimney Rock Animal Trap.
    These sites are interpreted as having had tundra habitat
nearby, most probably on mountain tops and exposed northern
slopes. Boreal forest covered much of the southern slopes
and protected areas. Subalpine meadows and open sagebrush
slopes completed the suite. Similar habitat can be found
today in the Northern Rockies, but present tundra areas lack
the continuity or near-continuity that must have been present
during the Pleistocene for _Dicrostonyx_ to spread so widely.
To cross as a near-continuous strip into the Southern Rockies, tundra habitat must have descended to about 6850 m ee or
below.
    Each of these sites is considered below in order, from
highest to lowest elevation, equator equivalent. Citations

following site elevations are the primary sources for the modern site data.

Complete faunal lists for these sites are given in Appendix 3.

Jaguar Cave, Lemhi Co., ID.   2255 m, 7017 m ee (Guilday and Adam, 1967).

This site lies at the mouth of a canyon opening into Birch Creek Valley. Currently, riparian, grassland, swampland, and sagebrush habitats interspersed with low pines occur near or below the site; presumably boreal forest occupies the nearby Beaverhead Mountains. Wisconsinan valley glaciers reached or nearly reached the cave site; deposition may have commenced at the beginning of Pinedale recession and ceased by the beginning of the Altithermal.

Guilday and Adam (1967) stressed the absence of woodland forms, but presence of Lepus americanus, Vulpes vulpes (red fox), Gulo gulo, Martes americana (martin), and Lynx rufus (bobcat) suggest forest was in the near vicinity.

The record of Vulpes cf. macrotis (kit fox) seems likely to be due to misidentification or intrusion, and the specimen should be reexamined critically. The Pleistocene distribution of the lesser red foxes indicates V. velox would be expected, but not V. macrotis (the former occurred about 200 m ee lower, but the nearest record elevationally of V. macrotis was over 1500 m ee lower). Vulpes macrotis occurs only south of the area today, while V. velox extends far to the north, although only east of the site.

Horned Owl Cave, Albany Co., WY.   2439 m, 6880 m ee (Guilday et al., 1967).

This site is 34 km northeast of Laramie on the western

slope of the Laramie Mountains. The Laramies rise to about 2600 m in this area, supporting scattered pines. The site was not sealed and much material was without provenience; thus Holocene material may have been included.

Presence of <u>Ochotona</u> <u>princeps</u> and <u>Oreamnos</u> cf. <u>americanus</u> is the basis for the tentative assignment of tundra habitat; both may appear in lower-elevation habitats.

Bell Cave, Albany Co., WY. 2379 m, 6820 m ee (Zeimens and Walker, 1974).

Bell Cave is in the Laramie Mountains directly across Wall Rock Canyon from Horned Owl Cave. Holocene material was present in addition to the Pleistocene fauna.

Chimney Rock Animal Trap, Larimer Co., CO. 2410 m, 6797 m ee (Hager, 1972).

Located 30 miles southwest of Laramie, WY, this site lies in sagebrush plains. Holocene material was present.

Watino, Alberta. Est. 600 m, est. 6540 m ee (Churcher and Wilson, 1979).

This site is west of Lesser Slave Lake on a tributary of the Peace River, more than 145 km east of the British Columbia provincial border. Prairie habitat is dominant now. The area became habitable by terrestrial mammals about 10,700 ^{14}C years ago, following local retreat of continental glaciation.

Presence of <u>Ovibos</u> cf. <u>moschatus</u> (muskox) indicates tundra was present. The other extant forms, <u>Cervus</u> <u>elaphus</u> (elk) and <u>Bison</u> <u>bison</u> (bison), reveal that grasses were present, but in themselves fail to indicate the vegetational context of the grasses. Churcher and Wilson (1979:75) suggested the climate "to have been slightly drier and possibly warmer, and the vegetation to have been more nearly that of open prairie, or aspen or poplar woodland rather than closed forest or coniferous forest." With the exception of the "possibly warmer," I would concur.

Cochran, Alberta. Est. 1067 m, est. 6524 m ee (Churcher, 1968).

The Cochrane faunal material came from terrace gravel deposits along the Bow River. Dates for the site indicate the area to have been free of ice for 3000 to 4000 years at the time of deposition.

Tentatively, the area is considered to have been tundra because of the presence of Rangifer, but presence of this taxon does not rule out open boreal forest. Other living taxa indicate an open habitat, however.

Granite Canyon No. 1, Juab Co., UT. 2070 m, 6350 m ee (Thompson and Mead, 1982).

Plant remains of Juniperus communis and Pinus flexilis in a woodrat midden indicate boreal forest had been present; Ochotona is indicated by the presence of fecal pellets in the midden. Although sagebrush also occurred in the midden, the species was not given in the publication.

Little Box Elder Cave, Converse Co., WY. 1676 m, 6224 m ee (Anderson, 1968).

This cave site is 29 km west of Douglas, WY, in the foothills of the Laramie Mountains. Elevations of >1870 m occur within about 1.5 km of the site. Bones occurred throughout the four layers present, becoming increasingly common in the lower levels; Pinus ponderosa needles were present in the

second and third layers. Different times appear to be represented by the four levels.

Although not clear from Anderson's (1968) report (pp. 7-8:
"The contacts between the layers are distinct." "wood rats .
. . are present in the cave, and the distribution of bone is
probably due to their activities."), probably some mixing of
Holocene and Pleistocene material has occurred. Distribution
of the fauna in terms of level was not given.

Spermophilus variegatus was north of its currently accepted range and some 750 m ee above its next highest late Wisconsinan occurrence. It may have been intrusive, representing an Altithermal increase in range. Long (1971) implied
that the occurrence of Cryptotis (least shrews) should be
dated from a range expansion during the climatic optimum
(Altithermal); it appears more likely, however, that present
restriction of Cryptotis to a range east and southeast of the
site is due to low precipitation rather than cold climate
(see Porter, 1978). Cryptotis seems more likely to have
spread westward during a pre-Altithermal time of greater
effective moisture, though perhaps only forced to elevations
as high as Bell Cave during the warm Altithermal. Porter's
(1978) data seem to preclude habitation of a peri-tundral
area, and thus presence of Cryptotis seems to emphasize the
probable time-transgressive nature of the site.

Propects Shelter, ca. 1400 m, ca. 6215 m ee (Chomko, 1978;
Mears, 1981) and Natural Trap Cave, Big Horn Co., WY. 1400
m, 6215 m ee (Martin and Gilbert, 1978).

These two sites are close to each other. Information
other than faunal occurrences is limited to Natural Trap
Cave. This site is in the so-called Juniper Breaks on the
northwestern flank of the Big Horn Mountains, northeast of

Lovell. Current vegetation is short grass/sagebrush steppe. Seven major and four minor strata occurred, with ^{14}C dates of 12,000 BP near the top of stratum 2 and 14,000 near the bottom of that stratum. Stratum 3 had dates of 17,620 near the top and 20,170 at the bottom. Stratum 3 was the major bone-bearing level. Stratum 4 was the lowest stratum with bone. Evidence indicates a former cone of ice and snow was present under the opening.

Martin and Gilbert (1978:113) reported that preliminary analysis by P. S. Wells of nearby woodrat middens indicated rare spruce occurrence in the vicinity; the top of the plateau around the cave "may have been open with a plant cover similar to that found in the high Alpine meadows of the modern Big Horns." Absence of permafrost evidence and of tundra forms other than Dicrostonyx led them to suggest a steppe tundra rather than true tundra. Prospects Shelter, however, included a muskox (Ovibos/Symbos) (Chomko, 1978).

Little Canyon Creek Cave, Washakie Co., WY.   Est. 1430 m, est. 6138 m ee (Frison and Walker, 1978; Mears, 1981).

The elevation of this site, located in the western foothills of the southern Big Horn Mountains, is a very rough estimate. It has not been given in the literature. Present vegetation is sagebrush/juniper. The fauna is older than 10,170 $\pm$ 250 (RL-641).

Frison and Walker (1978:200) indicated "a strong possibility . . . for the nearby presence of some form of arctic steppe/savanna conditions . . . ."

Smith Creek Cave; Smith Creek Cave No. 4; No. 5, White Pine Co., NV.   1950 m, 6123 m ee (Thompson and Mead, 1982; Mead et al., 1982).

These sites lie in a deep canyon on the eastern face of the Snake Range. The rough topography results in complex plant community distribution, with representation of xerophytes, Artemisia tridentata, woodland, and montane forest elements relatively near the cave.

The two woodrat middens (Smith Creek Cave No. 4 and No. 5) included fecal pellets of Ochotona. Plant remains included Pinus longaeva, Juniperus communis, and Artemisia sp. Dates were 12,235 ± 395 (GX-5863) and 13,340 ± 430 (A-2094).

Likely candidates for time spans earlier or later than that studied here are Ammospermophilus leucurus (white-tailed antelope squirrel), Perognathus cf. parvus (Great Basin pocket mouse), Microdipodops cf. megacephalus (dark kangaroo mouse), Dipodomys cf. ordi (Ord's kangaroo rat), Neotoma lepida (desert woodrat), and Panthera onca (jaguar). In addition, the identifications (both questioned) of Antrozous (pallid bats) and, particularly, †Capromeryx (diminutive pronghorns) seem out of place. The latter did not appear again until under 5000 m ee, some 1150 m ee lower than Smith Creek Cave, but appeared regularly from that point downward.

Empress, Alberta. Est. 640 m, est. 6097 m ee (Harington, 1978).

This site is in a terrace on the Red Deer River. The only extant taxon recovered was Rangifer sp., implying either tundra or boreal forest.

Hell Gap Site, Goshen Co., WY. 1525 m, 6073 m ee (Roberts, 1970).

This site is near the contact between the plains to the east and the foothills of the Laramie Mountains to the west. Riparian growth and Transition Life Zone vegetation (Pinus

ponderosa on slopes, sagebrush and grasses on the flats)
occur now near the site.

Roberts (1970:13) suggested "open Transition or Canadian
life zone woodland" with "patches of open meadow in the valley as well." However, the taxa on which he based the woodland were Marmota flaviventris (yellow-bellied marmot), Neotoma cinerea (bushy-tailed woodrat), Erethizon dorsatum
(porcupine), and Lepus sp. These are not limited to nor are
they necessarily typical of woodland growth.

Douglas, Converse Co., WY. 1468 m, 6069 m ee (Walker, 1982).

The site, in a river terrace, is presumed to be late Wisconsinan in age. Walker (1982) suggested the single specimen
of Ovibos moschatus indicated a steppe or steppe/savanna
environment.

Middle-elevation Savannah Zone

A second set of sites lacked evidence of tundra, but includes
sites between 5024 and 6063 m ee. The former is the lowest
elevation interpreted as having included boreal forest elements. The only sites within the span that lacked boreal
elements are those in areas lacking high relief: Dutton and
Selby on the plains; Isleta Cave No. 1 and No. 2 on the high,
relatively low-relief basin surface of north-central New Mexico; and Howell's Ridge Cave in an outrider of the low Little
Hatchet Mountains of New Mexico.

Evidence of former presence of good sagebrush growth
occurred to as low as the Isleta Caves (no evidence at the
Sheep Range middens or Dutton); below 5461 m ee, evidence of
sagebrush within this zone was limited to Howell's Ridge

Cave, where Centrocercus urophasianus and Lagurus curtatus
were found. However, since sagebrush is recorded as having
occurred below this zone to as low as 4578 m ee (Dark Canyon
Cave), it is possible that sagebrush occurred at all of the
sites.

Evidence of well developed grassland is sporatic, with
western sites particularly apt to lack such evidence.

Streamview No. 1, No. 2, White Pine Co., NV. 1860 m, 6033 m
ee (Thompson and Mead, 1982; Mead et al., 1982).

This shelter in Smith Creek Canyon is on a north-facing
slope; protected areas near the site support mesic conifers.

Fecal pellets record the past presence of Ochotona in
Streamview No. 1 at 11,010 ± 400 (A-2095), along with Pinus
flexilis, P. longaeva, and Artemisia sp.; and in Streamview
No. 2 at 17,350 ± 435 (GX-5866) with P. flexilis, Juniperus
communis, J. scopulorum, and Abies concolor.

Sheep Range, South Crest Midden, Clark Co., NV. 1990 m, 5896
m ee (Spaulding, 1977; Thompson and Mead, 1982).

This woodrat midden was at 1980 m (Spaulding, 1977) or
1990 m (Thompson and Mead, 1982) on an arid slope facing
southeast. Joshua-tree (Yucca brevifolia), snakeweed, mountain mahogany (Cercocarpus intricatus), and Artemisia tridentata occur near the site today.

Pinus flexilis, P. longaeva, Abies concolor, Artemisia
tridentata, Pinus monophylla, and Juniperus osteosperma are
among the midden plants recorded.

Wilson Butte Cave, Jerome Co., ID. Ca. 1311 m, ca. 5859 m ee
(Gruhn, 1961).

Wilson Butte Cave is located in the Snake River Plains.

At present, the immediate area is semi-arid, with precipitation of <25 cm per year. Artemisia tridentata and grasses
are the dominant plants on the butte. Modern fauna includes
Lepus californicus (black-tailed jack rabbit), Marmota flaviventris, and Neotoma cinerea.

The lower strata (E, D, and C) were believed by Gruhn
(1961) to have been deposited under cooler, moister conditions. Stratum D showed distortions interpreted to be the
effect of frost. The last extinct megafauna was in the middle of Stratum C (the faunal list given in Appendix 3 for the
site includes the fauna only through the middle of Stratum
C). Stratum C showed intrusive animal burrows.

Agate Basin Site, Niobrara Co., WY. 1158 m, 5813 m ee
(Mears, 1981).

This site is located on the plains of extreme eastern
Wyoming. Present vegetation is presumed to be prairie.

Garrison No. 1, No. 2, Millard Co., UT. 1640 m, 5813 m ee
(Thompson and Mead, 1982; Mead et al., 1982).

These middens were in a rock outcrop within a valley some
10 km east of the Snake Range of Nevada. Garrison No. 1 was
dated at 12,230 ± 180 (A-2312) and No. 2 at 13,480 ± 250
(A-2313). Pinus flexilis and Artemisia appeared in the midden deposits. Thompson and Mead (1982:42) suggested that in
Great Basin valleys, "sagebrush and other shrubs were present
on alluvial and lacustrine substrates during the Late Wisconsin," while rock outcrops supported subalpine conifers.

Atlatl Cave, San Juan Co., NM. 1910 m, 5762 m ee (Betancourt
and Van Devender, 1981).

Atlatl Cave, on the north wall of Chaco Canyon, had a

series of woodrat middens that included material dated from
10,600 ± 200 (A-2139) to 9460 ± 160 (A-2116). The middens
recorded communities apparently dominated by Pinus cf. flexilis, Pseudotsuga menziesii, and Juniperus scopulorum. Neither Pinus ponderosa nor P. edulis were present. Artemisia
tridentata-type occurred.

Gillespie (pers. comm.) recorded Lagurus curtatus from
Atlatl Cave with evidence to suggest that it dated from this
general time period or earlier.

Sheep Range, Flaherty Mesa No. 1, Clark Co., NV. 1770 m,
5676 m ee (Thompson and Mead, 1982).

This woodrat midden included both bones and fecal pellets
of Ochotona. Plant remains included Pinus flexilis, P. monophylla, and Juniperus osteosperma. There is a date of 20,380
± 340 (WSU-1862).

Dutton, Yuma Co., CO. 1255 m, 5482 m ee (R. W. Graham,
1981).

The Dutton site lies in the sand grass-sand sage association within the sand dune country of northeastern Colorado. Present faunal affinities tend to be to the east.

Graham (1981) considered the environment to have been one
of grassland. Although I agree that this was the dominant
habitat, presence of Urocyon may indicate riparian stringers
were in the area.

Isleta Cave No. 1, No. 2, Bernalillo Co., NM. 1716 m, 5461 m
ee (Harris and Findley, 1964).

The Isleta Caves are lava pit-traps on the generally low-
relief surface west of the Rio Grande Valley in north-central
New Mexico. Present vegetation consists of arid grassland

interspersed with shrubs such as four-winged saltbush (<u>Atriplex canescens</u>); a few <u>Juniperus monosperma</u> occur about lava outcrops. A few miles to the east, one of the northernmost stands of creosotebush (<u>Larrea divaricata</u>) in New Mexico persists on basaltic slopes.

These sites have been open from some time in the late Wisconsinan (two ^{14}C dates on camelid bones recently have been submitted for dating) to the present; both Pleistocene and Holocene material was entrapped. Original excavation notes from the early 1940's (University of New Mexico Anthropology Department) have been lost and subsequent collections (UTEP) have been from highly disturbed areas.

Many of the bones appear burned. This is believed to have been by natural means. At present, large amounts of plant debris (now mostly tumbleweeds) accumulate, particularly in Isleta No. 2; natural fires on the surface would ignite such debris.

Several taxa have been identified since the original publication, including cf. <u>Centrocercus urophasianus</u>, <u>Brachylagus idahoensis</u>, <u>Sylvilagus nuttalli</u> (Nuttall's cottontail), <u>Mustela</u> cf. <u>vison</u>, and †<u>Panthera leo atrox</u> (lion). Taxa previously listed to genus but now tentatively identified to species include <u>Microtus</u> cf. <u>pennsylvanicus</u> (meadow vole), †<u>Equus</u> cf. <u>conversidens</u> (Mexican ass), †<u>E.</u> cf. <u>niobrarensis</u> (Niobrara horse), †<u>Camelops</u> cf. <u>hesternus</u> (yesterday's camel), and †<u>Hemiauchenia</u> cf. <u>macrocephala</u> (large-headed llama).

From their state of preservation, some materials seem to have been deposited at a relatively late, probably Holocene, date. This was not adequately treated in the original report and several taxa have been assumed in the literature to be Pleistocene that almost certainly are not. These include most of the reptilian material identified: at least, <u>Cnemi-</u>

dophorus perplexus, Salvadora grahamiae, Pituophis catenifer
(=melanoleucus), and Lampropeltis getulus. Recently obtained
records of correspondence received from the University of New
Mexico Dept. of Anthropology indicate large tortoise remains
were recovered; efforts to locate these specimens have thus
far been unsuccessful.

Separating Holocene from Pleistocene mammals is more difficult. Certainly many of the specimens representing forms
now living in the area are Holocene in age; without further
study, however, it cannot be stated categorically that some
are not Pleistocene. The problem is under current study.

Although not provable at present, the following forms are
considered here as Holocene-only: Sylvilagus auduboni (desert cottontail), Lepus californicus, Spermophilus spilosoma
(spotted ground squirrel), Perognathus flavus (silky pocket
mouse), P. intermedius (rock pocket mouse), Onychomys torridus (southern grasshopper mouse), Vulpes macrotis, and Ovis
aries (domestic sheep).

Selby, Yuma Co., CO. 1173 m, 5453 m ee (R. W. Graham, 1981).
The Selby site is similar to the nearby Dutton site.

Upper Sloth Cave (C-08), Culberson Co., TX. 2000 m, 5425 m
ee (Logan and Black, 1979; Van Devender et al., 1977).
The Upper Sloth caves (three caves are designated as the
Upper Sloth caves: C-05, C-08, and C-09) in the southern
Guadalupe Mountains are located on the steep western face.
Current vegetation includes "a complex, high-elevation Chi-
huahuan desertscrub mixed with chaparral and grassland spe-
cies" (Van Devender et al., 1977:15). Two relict Pinus
edulis in protected areas are the only trees nearby.
Plant fossils from Upper Sloth Cave (C-08) and nearby Dust

Cave (C-09) included Picea sp., Juniperus communis, Pseudotsuga menziesii, Pinus strobiformis, and Pinus edulis (Van
Devender et al., 1977). This flora had dates from the Upper
Sloth caves of 13,000 ± 730 (A-1539) and 13,060 ± 280 (A-
1549).

The main faunal material (Logan and Black, 1979) came from
four 10-cm level units and was reported only in terms of
these levels. Unfortunately, the levels did not follow the
natural stratigraphy. The Pleistocene unit (uncontaminated?)
ran from an upper depth of 6 to 10 cm to approximately 40 cm;
therefore the faunal material from the 0 to 10-cm level may
include both Holocene and Pleistocene material. A radiocarbon date of 11,760 ± 610 (A-1519 according to Logan and
Black, 1979; A-1533 according to Van Devender et al., 1977)
on artiodactyl dung came from the 30 to 40-cm level. Plant
debris from above the dated layer indicated a lower-elevation
forest and woodland than did the approximately 13,000-year
old material from the woodrat midden reported above. Plants
included Pseudotsuga menziesii, Pinus edulis, Juniperus sp.,
and Quercus gambeli (Van Devender et al., 1977).

Peromyscus eremicus (cactus mouse) was identified from the
0 to 10 and the 10 to 20-cm levels. Identification was by
tooth characters that, unfortunately, also are typical of P.
crinitus (canyon mouse). The latter species apparently was
not considered, though reported tentatively from nearby Dry
Cave (Harris, 1970). Lower Sonoran Life Zone P. eremicus is
distinctly out of place, whereas P. crinitus more closely
fits the suggested habitats. It also fits the common pattern
in the Guadalupe Mountains area, where Pleistocene mammals
with affinities to the northwest were common.

Logan and Black (1979:156) postulated "that approximately
11,000 years ago the climatic conditions in the southern

Guadalupe Mountains may have been similar to the climatic conditions existing in the Black Hills of South Dakota today" on the basis of the sympatric occurrence of Sorex cinereus (masked shrew), Cryptotis parva (least shrew), Marmota flaviventris, and Neotoma cinerea in the Black Hills today.

Lower Sloth Cave (C-05), Culberson Co., TX. Ca. 2000 m, ca. 5424 m ee (Logan, 1977).

Lower Sloth Cave is located at the intersection of a steep talus slope and the base of a cliff. The present vegetation is the same as at the nearby Upper Sloth Cave (Upper Sloth, Lower Sloth, and Dust caves lie within 400 m of one another). Fossil plant remains included ? Pinus edulis and Juniperus sp.

Six test trenches were dug in 10-cm increments.

Among the species identified, the identification of Spermophilus spilosoma is suspect because of its present ecology (generally no higher than Upper Sonoran grasslands) but, more particularly, because of the likelihood of confusion with the similar S. tridecemlineatus (thirteen-lined ground squirrel). The latter, a more mesic species that also occurs in the Southwest today, would be the expected species.

The occurrence of Onychomys torridus likewise is surprising. Today, this species is a Lower Sonoran form not expected even as high as Upper Sonoran grasslands. Confirmation of the identification and Pleistocene age would pose interesting questions as to the ecological relationships of this species and O. leucogaster.

Dust Cave (C-09), Culberson Co., TX. 2000 m, 5424 m ee (Logan, MS; Van Devender et al., 1977).

Pleistocene plant species (Van Devender et al., 1977)

essentially were the same as identified from Upper Sloth Cave.

Boreal and open coniferous forest, grassland, and woodland habitats appear to have been represented by the faunal remains.

Mescal Cave, San Bernardino Co., CA. 1550 m, 5349 m ee (Brattstrom, 1958).

This site is in the relatively low Mescal Range; site information is unavailable.

Too few taxa were present for comfortable interpretation, but most likely boreal forest elements accompanied by open patches of grass were nearby.

Howell's Ridge Cave, Grant Co., NM. 1675 m, 5099 m ee (Smartt, 1977; Van Devender and Worthington, 1977; UTEP).

This site is a vertical chimney in the side of a low limestone ridge of the Little Hatchet Mountains. The chimney opens laterally to near the surface of the sediments and probably did not act as a pit trap. Upper portions of the chimney are ideal for raptor nesting and perching. The present vegetation is ecotonal between Lower and Upper Sonoran life zones. A playa lies in the Playas Valley a few kilometers to the west.

Van Devender and Worthington (1977) interpreted the sediments in their test pit as undisturbed; Van Devender and Wiseman (1977:18) reported three of their six ^{14}C dates that were based on plant material as questionable and probably "due to contamination by younger materials reworked into these levels." Nevertheless, both publications basically interpreted the faunal material as if it were undisturbed.

An earlier, nearby test pit was excavated by myself and

Smartt (Smartt, 1977). Our belief at the time was that the
deposits were disturbed; it is unlikely that our pit, close
to that of Van Devender and Worthington's, differed greatly
in degree of disturbance. This is bourne out by Van Devender
and Wiseman's dates (Table 5) and by several lines of faunal
evidence. Also, the Gymnogyps californianus bone dated at
13,640 ± 220 (A-1557) was from our pit, which reached a depth
of only 1.34 m; however, Van Devender and Worthington (1977)
estimated a date of only about 11,500 from their lowest level
of 1.70-1.80 m. Likewise, Smartt (1977) identified four species of Microtus from within the upper 1.34 m of fill; this
implies more complex conditions than the interpretation of
"more mesic grass habitats in Playas Valley" (Van Devender
and Wiseman, 1977:20).

Also, if the dampness of the lower levels that destroyed
the original plant matter below about 90 cm (Van Devender and
Wiseman, 1977) was in part a function of depth of burial (as
it appears to me), then continual mixing of upper layers during deposition would have introduced older organics (before
burial to a depth conducive to decay) to the upper levels.
Thus the upper dated levels likely showed ages somewhat too
old because of mixing of some old material with young; the
deeper levels would appear much too young because only relatively recently mixed in younger material would have survived.

All of these features are consistent with mixing of
Pleistocene-age material with that of the Holocene. Van
Devender and Worthington recorded increasing ratios of mesic
herptiles with depth; moderate mixing during deposition would
not obliterate evidence of climatic change, but would stretch
out changes temporally so that a change in fauna would appear
to commence lower in the column than actually should be the

TABLE 5. Radiocarbon dates from Howell's Ridge Cave,
Grant Co., NM (Van Devender and Wiseman, 1977). Level
data on A-1557 by author.

Level (cm)	Date	Laboratory No.
40-50	2470 ± 120	A-1574
70-80	3330 ± 170	A-1354
70-90	3910 ± 80	A-1619
110-120	6700 ± 325	A-1429 ± A-1430
140-150	930 ± 140	A-1659
150-160	1170 ± 140	A-1658
150-160	4690 ± 800	A-1575
<134	13,640 ± 220	A-1557

case and terminate at a level too high. Sharp changes would
be blurred into gradual changes.

I interpret the Pleistocene material to represent primarily grassland with some sagebrush; sheltered slopes likely
supported open coniferous forest or woodland.

Muskox Cave, Eddy Co., NM. 1600 m, 5024 m ee (Logan, 1981).

This cave is a sinkhole on the eastern side of the Guadalupe Mountains of southeastern New Mexico. Present vegetation is a mixture of Chihuahuan Desert forms with grassland
and chaparral types. Radiocarbon dates on bone collagen run
from 25,500 ± 1100 to 18,140 ± 200.

Logan (1981:159) interpreted the area (partly based on the
paleobotany from the Upper Sloth caves and Williams Cave) as
"spruce-fir forest with open grassy meadows, probably with a
small permanent stream along the valley floor."

I would suggest open coniferous forest with a decidedly
boreal aspect (perhaps along drainageways) rather than pure
boreal forest (which may not have been implied by Logan).

Along with the grassland may have been woodland elements, too.

Several taxa do not appear to fit either interpretations and may represent other than full-glacial conditions: Onychomys torridus and Neotoma micropus, particularly.

Hermit's Cave, Eddy Co., NM. Est. 1600 m, est. 5021 m ee (Ferdon, 1946; Schultz et al., 1970).

This site presumably lies in the Lower Sonoran/Upper Sonoran ecotone, or perhaps in woodland today, though no information is given in the primary sources. The site lies near the top of an approximately 20-m talus slope leading to the stream bed of Last Chance Canyon. Rugged cliffs lie above, with dissected terrain at their tops leading to a flatter rolling surface. The description is unsatisfactory for determination of distances to that surface, but the site is assigned to the foothills category for analysis.

Dates on wood from a hearth were 12,900 ± 350 and 11,850 ± 350.

Again, the reported fauna is too scanty for proper interpretation, but probably open coniferous forest, perhaps with boreal elements, and possibly grassland were in the vicinity.

Sagebrush Steppe-woodland Zone

Full-glacial vegetation at elevations (ee) below Muskox Cave and above 4339 m ee were characterized primarily by woodland growth, with grasses common. Sagebrush habitat is indicated as having been present at some sites. Strictly boreal elements seem to have been absent.

Blackwater Draw Sites, Roosevelt Co., NM. Ca. 1280 m, ca.
4972 m ee (Lundelius, 1972; Slaughter, 1975).

A series of sites along Blackwater Draw, in the grasslands
of extreme eastern New Mexico, have produced late Wisconsinan
faunas. The major faunas were from Blackwater Locality No.
1, and included material from the gray sands (Lundelius,
1972), believed by Haynes and Agogino (1966) to be more than
12,000 years old and possibly more than 16,000. At the same
site, the Brown Sand Wedge local fauna (Slaughter, 1975) probably was close to the age of 11,170 ± 360 (A-481) recorded
nearby (Haynes, 1975).

The faunas were basically grassland faunas, with the Brown
Sand Wedge showing a hint of woodland or open coniferous forest in the form of Peromyscus truei and Microtus cf. mexicanus. However, both of these forms are not only in difficult
genera for identification, but the records also were based
only on single teeth.

In addition to the more or less standard habitat(s), a
distinct eastern riparian element was indicated by a non-
coniferous forest Sciurus (identified as the Arizona gray
squirrel, S. cf. arizonensis, but on zoogeographic grounds,
much more likely to have been the eastern gray squirrel, S.
carolinensis), the opossum (Didelphis virginianus), and raccoon (Procyon lotor). In addition, the extinct armadillo,
†Dasypus bellus, and the peccary, †Platygonus, were forms
absent from most of the Inland Division, but apparently man-
aged to reach Blackwater Draw and the more northern Agate
Basin Site from the east.

The Brown Sand Wedge is one of the few late Wisconsinan
sites within our area to have had Sigmodon (cotton rat) present, implying somewhat warmer winter conditions or a longer

growing season (see species account). This may record warming temperatures at the end of the Pleistocene proper.

Papago Springs Cave, Santa Cruz Co., AZ. 1586 m, 4957 m ee
(Skinner, 1942).

Present vegetation of the site in the Canelo Hills of
southern Arizona is unreported, but presumably open woodland
with a grassy understory. The Pleistocene fauna indicates
there was a relatively mesic woodland with good grasses, the
major difference from today being the presence of Microtus ?
mexicanus and the extinct forms.

Williams Cave, Culberson Co., TX. 1495 m, 4919 m ee (Ayer,
1936; Van Devender et al., 1977).

This cave is in Texas at the southern end of the Guadalupe
Mountains. Chihuahuan Desert vegetation surrounds the site,
with a few junipers at the same elevation about 1 km away on
a particularly favorable substratum.

A woodrat midden dated at 12,040 ± 210 BP (A-1540) contained Pinus edulis, Juniperus sp., and Quercus sp., among
others. Thus a woodland habitat is indicated. The reported
fauna was a mixture of Pleistocene and Holocene material.

Burnet Cave, Eddy Co., NM. Est. 1435 m, est. 4910 m ee
(Howard, 1932; Schultz and Howard, 1935).

This site was one of the first reported from the Guadalupe
Mountains area of southeastern New Mexico. Although reported
by Howard (1932) as at an elevation of about 1400 m, the
described area is of higher elevation, estimated here at ca.
1435 m, but possibly somewhat higher. Amadeo Rea has suggested (in litt.) that Rocky Arroyo Cave, treated separately

by a number of authors, is synonymous with Burnet Cave, and I concur.

The modern vegetation about the site apparently has not been described in the literature. The scanty information I have available indicates an ecotonal situation between Upper and Lower Sonoran life zones, similar to that described below for Dry Cave.

With a few exceptions, stratigraphic data were not published and almost certainly some of the recorded forms were Holocene. Upper levels contained Christian-era archaeological remains.

A solid-carbon date of 7432 ± 300 was obtained from material associated with extinct fauna and with a Clovis-type point; the date was rejected on archaeological grounds by Hester (1964) and, based on faunal grounds, I agree; the probable date at that level (2.44 to 2.75 m) was between 11,000 and 11,500 BP.

I have examined some of the specimens (equids and leporids) on loan from the Philadelphia Academy of Science. Although corrected identifications are given in the taxonomic lists, it seems worth emphasizing that the specimen identified as Lepus alleni (antelope jack rabbit) is not identifiable to species on present knowledge (nor is the identification of Lepus townsendi, though that taxon seems likely in light of other faunas of the area). Likewise, the specimen identified as Sylvilagus floridanus (eastern cottontail) could be either S. floridanus or S. nuttalli; the S. auduboni identified, however, probably is that species, though S. floridanus cannot be ruled out entirely.

My tentative identification of †Equus tau is based on partially digested material, perhaps indicating that it did not occur in the local fauna, but was carried to the site from a

considerable distance inside the digestive tract of a carni-
vore. The semi-digested condition makes sure identification
difficult.

I have elsewhere suggested (Harris, 1981) that Schultz and
Howard's identification of Neotoma lepida probably referred
to the present N. stephensi because of nomenclatural changes.
Since that time, a small woodrat that definitely was not N.
stephensi has been studied; this rat, from Dry Cave, has been
identified as N. ? goldmani. The same taxon has been identified to the west, at Anthony Cave, and thus it seems likely
that the Burnet Cave specimen represented this taxon.

Several habitats seem to have been represented at Burnet
Cave, including grassland, some pinyon-juniper woodland, and
brushy type habitat.

Shelter Cave, Dona Ana Co., NM. 1475 m, 4899 m ee (Smartt,
1977; Thompson et al., 1980; Van Devender and Everitt, 1977).

The present flora is northern Chihuahuan Desert desert-
scrub; the area is relatively xeric for its elevation. The
site is a deep rock shelter on the western slope of a low,
limestone peak (Bishops Cap).

A series of radiocarbon dates and woodrat middens were
obtained from the site and its vicinity (Thompson et al.,
1980). Three sloth dung dates ranged from 11,130 ± 370
(A-1878) to 12,430 ± 250 (A-1880). Five dates on desert tortoise (Gopherus agassizi) scutes ranged from 11,130 ± 500
(A-2141) to 12,520 ± 200. Plants from older woodrat middens
dated from >26,000 (A-1753, A-1766) to 31,250 ± 2200 (A-
2140). Post-Pleistocene fill also was present.

The woodrat midden material (Thompson et al., 1980:362)
indicated "a relatively xeric juniper woodland," with sparse
Pinus edulis. Artemisia tridentata-type was recorded, but

was uncommon. The only marker species of mammal, Microtus
cf. montanus (montane vole), could have existed in this habitat if abundant grass was available.

Hueco Tanks No. 1, El Paso or Hudspeth Co., TX. 1420 m, 4844
m ee (Van Devender and Riskind, 1970).

   Unlike the preceding site, the Hueco Tanks area of extreme
western Trans-Pecos Texas is relatively mesic for its elevation, yet today supports mostly Lower Sonoran desert vegetation. A few higher-zone plants, such as Junperus erythrocarpa, occur in favored localities.

   The midden plants from this site were dated at 13,500 ±
250 (A-1624); a nearby site, Picture Cave No. 1D, dated at
12,030 ± 210 (A-1699). Van Devender and Riskind (1970:137)
interpreted the plant record as indicating "a relatively
mesic version of . . . pinyon-juniper woodland 13,500 radiocarbon years ago." The single mammalian fossil (Microtus
sp.) indicated good stands of grass were present, at least
locally.

Conkling Cavern, Dona Ana Co., NM. 1399 m, 4823 m ee (Conkling, 1932; Smartt, 1977).

   Conkling Cavern lies near Shelter Cave, but on the eastern
side of Bishops Cap; the present vegetation is Chihuahuan
Desert type.

   The vegetative interpretation based on the fauna was that
the area was one primarily of pinyon-juniper woodland, with
much grass. Brushy growth may well have been present.

Stanton Cave, Coconino Co., AZ. 927 m, 4779 m ee (Euhler,
1978).

   The present vegetation is sparse, with the principal

plants outside the cave consisting of grasses and various species of cacti. Little information was supplied by the Pleistocene fauna, but presence of sagebrush was indicated by the occurrence of <u>Centrocercus urophasianus</u>. Euler (1978) quoted a report by Hevly indicating that seed remains suggested a juniper woodland was present when the 40-cm level was deposited, somewhat earlier than ca. 10,760 BP. Euler (1978:156) also quoted a personal communication from Martin to the effect that the pollen from below 30 cm suggested "an environment in which Great Basin sagebrush occurred closer to the cave than it does at present, . . ."

Dry Cave Sites, Eddy Co., NM. 1280 m, 4758 m ee (Harris, 1970; UTEP).

Dry Cave, in rolling limestone country about half way between the Pecos Valley to the east and the Guadalupe Mountains to the west, lies in an ecotone between Lower Sonoran and Upper Sonoran life zones. It is an extensive cavern system with the present entrance a sink; additional sinks opened to the surface during the Pleistocene, but now are clogged.

A number of sites have been excavated within the cave, varying from Holocene to >33,000 radiocarbon years BP. Because of the time depth represented, the Dry Cave sites will be considered in a separate section further on.

Overall, the latest-Wisconsinan sites from Dry Cave indicated that the habitats centered on grassland and sagebrush, with pinyon-juniper woodland also represented.

Dark Canyon Cave, Eddy Co., NM. 1100 m, 4578 m ee (Howard, 1971; UTEP).

Dark Canyon Cave is the lowest of the Guadalupe Mountains

area sites. It is a large shelter in a low range of hills
about 155 m above the flood plain of the Pecos Valley a few
miles to the east.

No dates are available and most of the mammalian fauna
(the bulk of which is in the Texas Memorial Museum collections, but some of which is at UTEP) is unpublished. The
fauna thus far appears to indicate a pinyon-juniper woodland
with sagebrush and grass elements, but some forms were more
typical of higher zones. One tentatively identified mammal
(Tayassu, peccary) is of distinctly southern affinities,
currently reaching its northern limit in the same region.
Several birds also were of probable southern relationships
(Howard, 1971).

Tule Springs, E-1, Clark Co., NV. 703 m, 4555 m ee (Mawby,
1967; Mehringer, 1965).

Present vegetation of this site near Las Vegas is lower
Mojave Desert type. The site is in the open, relatively flat
Las Vegas Valley. Unit E-1 deposition started a bit earlier
than 13,000 and continued to slightly later than 11,500 BP
(Haynes, 1967).

Brachylagus (identification queried) is an indicator of
sagebrush habitat, and fits well with Mehringer's (1965:181)
pollen analysis that indicated the vegetation was "similar to
the juniper-sagebrush of northern Nevada today." Also, there
was "little doubt that junipers were abundant on the valley
floor in the immediate vicinity of the spring, at least 1000
meters lower than their present distribution in the Charleston Mountains." Microtus cf. californicus (California vole)
indicates grassy meadows were present, but the elevational
range of the species is great. Overall, the small mammal
fauna was too scanty to add appreciably to the pollen record.

Steppe-woodland Zone

Several sites lacked definite floristic and faunal evidence of Artemisia tridentata, but apparently supported grassy woodland growth.

Vulture Cave, Mohave Co., AZ.   645 m, 4497 m ee (Mead and Phillips, 1981).

This cave lies well within the Grand Canyon, surrounded by hot desert species of plants. Fossil flora and fauna were both associated predominantly with woodrat middens, which ranged in age from about 1160 to about 33,600 radiocarbon years BP (faunal elements listed here fell between 23,000 and 11,000 BP).

Middens dated between 23,000 and 15,000 BP indicated a xerophilous woodland dominated by Juniperus and Atriplex confertifolia, with blackbrush (Coleogyne ramosissima) and snowberry (Symphoricarpos sp.) present, as were several desert species still living in the area.

Middens from between 15,000 and 11,500 also indicated a xeric woodland, but single-leaf ash (Fraxinus anomala) was present and blackbrush dropped out early.

The faunal material added little. Microtus implied that fair grass was nearby, and the occurrence of Neotoma lepida and/or N. stephensi likely means woodland was in the area.

Gypsum Cave, Clark Co., NV.   610 m (elev. after Harrington, 1933; Kurten and Anderson, 1980, give as 454 m), 4462 m ee (Harrington, 1933; Mehringer, 1967).

This site lies in the hot desert of southern Nevada.

Dates of 8527 ± 250 (C-222) and 10,455 ± 340, both solid-
carbon dates, were not accepted by Mehringer (1967) and
others; Long and Martin (1974), however, noted dates obtained
by modern techniques of 11,690 ± 250 (IJ-452) and 11,360 ±
260 (A-1202) on sloth dung. Thompson et al. (1980) reported
two additional recent dates on sloth dung of 33,910 ± 3720
(A-1609) and 21,470 ± 760 (A-1611). Joshua tree (Yucca brev-
ifolius), a high elevation Mojave Desert plant, formed a
major element in the sloth diet (Laudermilk and Munz, 1934).

Only one taxon, Lepus townsendi, is of value for environmental reconstruction at this point. If the identification
is correct, a considerably cooler and more mesic environment
was present; in the Great Basin context, probably sagebrush
was represented.

Schuiling Cave, San Bernardino Co., CA. 658 m, 4403 m ee
(Downs et al., 1959).

This site, a deep shelter in the low, arid Newberry Mountains of the Mojave Desert, probably had at least its lower
sediments laid down during a period of stream valley aggrada-
tion brought on by the formation of a nearby pluvial lake.

The nearby lake also was indicated by the number of water
fowl in the deposits. The only useful mammalian fauna indicated probable woodland was in the area.

Rampart Cave, Mohave Co., AZ. 525 m, 4377 m ee (Mead, 1981;
Wilson, 1942).

This cave, within the Grand Canyon near the head of Lake
Mead, is well known for its ground sloth dung deposits. However, woodrat middens, unconsolidated plant materials, and a
limited vertebrate fauna together make this site one of great
importance in Pleistocene paleoecology.

Based on the fossil plant record, the local vegetation, now of hot desert species, was like that of nearby Vulture Cave--Juniperus-Fraxinus woodland with much Atriplex confertifolia--during the late Wisconsinan.

Numerous radiocarbon dates are available (Long and Martin, 1974), ranging from the 11th millenium BP to >40,000 BP.

The faunal evidence agrees with the floral in suggesting a woodland habitat was present.

Steppe Zone

Two sites, Jimenez Cave and Ventana Cave, are considered to have possibly lacked woodland, the major vegetation type having been grassland.

Jimenez Cave, Chihuahua.   1450 m, 4339 m ee (pers. comm., H. J. Messing; UTEP).

This cave site lies in southern Chihuahua, Mexico, in a group of low mountains. Playa lakes are nearby. Vegetation is hot Chihuahuan Desert type.

Dating of the deposits is uncertain, and most of the material was recovered from spoil discarded by presumed treasure hunters. The general character suggests it falls into our time span, but earlier times cannot be ruled out nor can contamination by Holocene material. It is included here despite the uncertainty because it is the only site from the entire state of Chihuahua and from well within the Chihuahuan Desert rather than on the northern fringes.

An aquatic bird fauna attests to the presence, at least seasonally, of lakes.

A healthy woodrat fauna included the nearly ubiquitous N.

albigula (white-throated woodrat), large rats representing N. micropus (southern plains woodrat) and probably N. floridana (eastern woodrat), N. lepida, and a taxon most similar to N. cinerea. N. floridana now occurs far to the north in the Central Great Plains, swinging south to central Texas and on to the east. N. micropus occurs now in the area. Jimenez Cave is far southeast of the present range of N. lepida, whose ecological distribution is from hot deserts into Upper Sonoran juniper woodland. Two specimens of woodrats are similar to N. cinerea, but differ enough to make identification uncertain; if N. cinerea, then likely woodlands were present. This site will be considered within the Steppe Zone for purposes of analysis.

Well developed grassland certainly was present, with presence of both southern (Sigmodon) and northern (Microtus pennsylvanicus) grassland creatures.

Ventana Cave, Pima Co., AZ. 750 m, 4228 m ee (Bryan, 1950; Colbert, 1950; Haury, 1950).

This site lies at the southern end of the Castle Mountains, in the Sonoran Desert of Arizona. Two levels contained extinct fauna, the conglomerate unit and the overlying volcanic debris layer. The latter was dated at 11,300 ± 1200 BP, consistent with the recovery of a Clovis-like point from the level (Mehringer, 1967).

The most notable taxon present was that identified as Callospermophilus lateralis (golden-mantled ground squirrel), now recognized as Spermophilus (C.) lateralis (Colbert, 1950). Mehringer and Lane reexamined the specimen (a lower jaw) and felt that a definite identification could not be made on it; however, three paleontologists who also examined

it at that time agreed with the identification to the subgenus Callospermophilus.

This squirrel today thrives in meadow and forest-edge situations in coniferous forests. Ecologically, it fits neither
the elevation ee nor the other extant members of the fauna.
Thus it seems likely a misidentification or, in view of the
talent that has examined it, possibly an example of long
range transport within the digestive system of a raptorial
bird. The possibility cannot be ignored, of course, that
some combination of environmental factors allowed spread into
vegetational types now avoided.

The other extralimital extant form, Cynomys ludovicianus,
is indicative of grassland rather than the desert vegetation
that is currently around Ventana Cave. However, as Mehringer
(1967) emphasized, it also occurred in most levels above the
volcanic debris; Colbert (1950:156) noted its occurrence in
the site as "highly spotty and infrequent." Sufficient
grassland to support small numbers may have lingered on to
historic or near-historic times.

Discussion

In attempting to reconstruct faunal makeup and vegetative
conditions from the types of evidence used here, a number of
weaknesses should be noted. One has been seen earlier in
that attempts to cancel the effects of latitude and elevation
may distort relationships. If the figure of 107 m per degree
of latitude is greatly inaccurate for the late Wisconsinan
times or the actual corrective figure for latitude varied
greatly with latitude itself or from east to west, then the
altitudinal order of sites as used here would be inaccurate.

However, it seems obvious that an imperfect correction is preferable to use of raw elevational data regardless of latitude and also that the results are consistent with usable accuracy.

A second source of weakness lies in the nature of the evidence itself. If many sites encompassed times of radical climatic change but are interpreted as being typical of stable, full-glacial times, then the faunas would appear to have been larger and more complex than actually was the case at any one point in time. Available dates are too few for most sites to determine the purity and some sites are known to have included some Holocene material (e.g., the Isleta Caves). Corrections for obvious error of this type have been made, but most such are uncorrectable at this time. The fluctuations at a single site (Dry Cave) will be considered later to give some insight into the problem.

Mis-identifications, always a factor with fossil material, further blur results as do the inequalities in site placement and faunal sizes. Inadequate knowledge of the parameters governing modern forms further complicates interpretation.

Nevertheless, a remarkably coherent pattern emerges when sites are plotted elevationally ee with the vegetative types as indicated by faunal and midden evidence (Table 4).

In Table 4, vegetative zones are assumed to have extended continuously from the lowest to the highest of those sites that showed good evidence for presence of the particular vegetative type (an exception is with midden evidence of woodland plants, since woodland elements frequently are minor aspects of higher-zone vegetation types). Thus, for example, the Northern Highland Zone is assumed to have extended from 6069 m ee to 7017 m ee even though not all sites showed evidence of tundra. Some such sites probably did not have

tundra near them, but in many cases absence of evidence
should be because of sampling error. Some of the sites which
might, by the elevation criterion, be expected to have been
within the zone are geographically excluded from any possibility of having possessed a tundra fauna by lowland areas
(e.g., Smith Creek Cave). Vegetative types considered present are minimal because criteria for acceptance of presence
generally require the least number of types consistent with
the fauna. For example, presence of a boreal marker species
plus the presence of Microtus pennsylvanicus, which occurs
both in minor grassland-sedge habitat within boreal forest
and in hydrosere conditions at lower elevations, would result
in scoring only boreal habitat, although another vegetative
type might in fact have been present. Thus any zone may actually have extended lower (or higher) than interpreted here.

When sites indicating that tundra or tundra-like vegetation was present are inspected, there is a high correlation
with evidence of boreal, sagebrush, and grassland elements.
Even if some of this is due to climatic fluctuations, all
vegetative types probably occurred nearby. I interpret the
situation as one in which all four vegetative types (and
their associated faunas) occurred in restricted areas.

Tundra, then, appears to have been restricted in extent.
In no site except those having an obviously inadequate fauna
was a tundra fauna unaccompanied by indicators of least two
other vegetative types (Watino and Douglas are considered too
limited to count, but, in view of the limited topographic relief of these sites, may have been steppe-tundra only). Tundra likely was limited to crests and northern slopes around
those sites located in mountains and foothills.

Boreal, sagebrush, and grassland elements not only were
associated regularly with each other in sites showing tundra

elements, but also occurred together below the tundra zone
to as low as 5700 to 5800 m ee (many of the sites between
6033 and 5762 m ee are somewhat equivocal, being midden sites
or sites with few faunal elements). The strength of the
association is considered so strong that the concept of a
boreal forest-sagebrush-grassland community, with each element being strongly represented, seems defensible. This
association is here designated as the Middle-elevation
Savannah.

Most of the sites with respectable numbers of faunal remains included a healthy grassland grazer component by usual
interpretation. These grazers commonly included <u>Ovis</u> <u>canad-
densis</u> (mountain sheep) and <u>Antilocapra</u> <u>americana</u> (pronghorn)
among living grazers as well as †<u>Camelops</u>, †<u>Equus</u>, and <u>Bison</u>.
Of these, <u>Ovis</u> normally requires basically open country
(Geist, 1971), but living <u>Antilocapra</u> and <u>Bison</u> occur or
occurred in areas of open woodland or mountain parks, respectively.

The near-ubiquity of grazers suggests that much of the
terrain was predominantly non-forested, though possibly with
scattered trees, with a closed or nearly-closed canopy limited to sheltered slopes and rugged terrain.

The microfauna tends to bear this out with such usually
open-area or forest-edge taxa as <u>Spermophilus</u> (<u>S</u>. <u>lateralis</u>,
<u>S</u>. <u>richardsoni</u>, <u>S</u>. <u>tridecemlineatus</u>, etc.), <u>Cryptotis</u> <u>parva</u>,
<u>Lagurus</u> <u>curtatus</u>, <u>Thomomys</u> <u>talpoides</u> (northern pocket goph-
er), <u>Brachylagus</u> <u>idahoensis</u>, <u>Lepus</u> <u>townsendi</u>, and <u>Marmota</u>
<u>flaviventris</u> common. More typically forest or brush dwell-
ers, such as <u>Clethrionomys</u> <u>gapperi</u> (red-backed vole), <u>Martes</u>
<u>americana</u>, forest cottontails, and <u>Lepus</u> <u>americanus</u> were rare
as were tree squirrels.

†<u>Mammut</u> (mastodon), considered a browser, perhaps particu-

larly on spruce, appeared only in the southern mountains (New Mexico). †Mammuthus (mammoth), a grazer, didn't appear until elevation decreased below about 6200 m ee in our sample.

Sagebrush appears to have dropped out as a ubiquitous member of this complex below about 5700-5800 m ee; below this elevation, to about Muskox Cave (5024 m ee), there was either evidence of sagebrush or grassland, but not both, at any given site.

Through this range (and, with the exception of Dry and Dark Canyon caves, to the lowest elevations of sagebrush), sagebrush appears to have been limited to sites near or west of the Rio Grande. This may be an artifact of sampling error or reflect climatic differences. Wells (1979) particularly has noted increased precipitation to the east; eastern sites near mountain masses may have picked up sufficient year-around moisture to allow grasses to have outcompeted sagebrush. Farther from the mountains, as at Dry Cave, precipitation patterns may have allowed coexistence.

Grasses maintained their association with boreal forest down to the lower elevational limit of the latter at Muskox Cave.

Several exceptions appeared to the general picture. Boreal forest sites lacking grassland faunal evidence included most midden sites, where sampling error is high due to low faunal content. Likely real, rather than due to small samples, was the absence of boreal forest evidence at five sites within the elevational range of 5024-7017 m ee. These were Hell Gap, Dutton, Isleta No. 1 and No. 2, and Selby. All are in areas of low relief. Boreal trees, if present, probably were too scattered to support a boreal fauna.

Evidence of boreal elements in the pollen record to the east (e.g., Oldfield and Schoenwetter, 1975) and the evidence

of modern disjunct populations in such places as the Black
Hills suggest continuous corridors in the late Wisconsinan.
Such connections may have been in other places or times than
represented here, or the boreal trees and their associates
may have been limited to the protected slopes of major river
valleys until they encountered sufficiently increased precipitation to the east to allow expansion onto the less protected interfluves. If an open parkland was represented,
then the waxing and waning of somewhat thicker than average
stands may have allowed such creatures as Tamiasciurus hudsonicus to cross a generally low-density parkland.

Below about 5700-5800 m ee, sagebrush continued to appear
sporatically. Occurrence of Lagurus curtatus at Dry Cave
suggests that sagebrush has occurred along every portion of
some strip between its present range and Dry Cave. This did
not have to be a continuous, synchronous strip; at any one
point in time, gaps may have separated patches of suitable
habitat, these patches migrating through time. Absence of
sagebrush indications at sites such as the Upper Sloth Caves
probably was a result of edaphic and topographic controls
rather than climatic, but as suggested earlier, may have been
the result of orographic precipitation patterns in the mountainous areas.

Most sites lower than the elevations where boreal elements
usually were accompanied by sagebrush were in areas of very
rugged topography, probably unsuitable for the larger stands
of sagebrush necessary to support Lagurus, Brachylagus, and
Centrocercus. Lack of detailed topographic information for
many sites in the literature makes this hard to test.

Occurrence of sagebrush below the elevation ee of Dark
Canyon Cave is uncertain. Tule Springs should have supported
such by the reconstruction of Mehringer (1965); the queried

identification of Brachylagus would tend to uphold this.
Identification of Lepus townsendi at Gypsum Cave, if correct,
would suggest sagebrush or sagebrush-grassland occurred as
low as 4461 m ee. The latter site is not included in the
Sagebrush Steppe-woodland Zone, however, for purposes of
interpretation.

Grassland of some type extended throughout to the lowest
elevations available, judging from a combination of extant
and extinct taxa. Throughout most of this span to as low as
Rampart Cave (4377 m ee) and also extending well into the
boreal complex, grasses were associated with woodland components. The upper elevational limit of the woodland is not
clear. Some elements appeared in middens as high as 5896 m
ee (Sheep Range-South Crest Midden), but satisfactory evidence of its occurrence as a significant form didn't appear
above the elevation of about 5424 m ee (Upper Sloth Caves).

Likely, woodland elements were scattered until about 5450
m ee, but in most lower sites down to Rampart Cave, formed
more or less coherent, though open, stands. Through various
parts of this span, boreal forest, sagebrush, and grasses
coexisted with it.

In the lower areas of boreal forest, woodland elements
probably inhabited shallow-soiled, arid exposures fringing or
separate from the boreal elements, but possibly interspersed
with open coniferous-forest trees. In local areas unsuitable for the latter, a pattern of sagebrush, grassland, and
woodland probably was present, continuing into lower elevations.

As noted above, grassland was associated with woodland
throughout the latter's span. This likely was in the form of
an open grassland with scattered woodland trees that became
more common in areas of shallow soils and rock outcrops.

Continuation of the open-country Ovis canadensis throughout
suggests the woodland did not approach (at least over extensive areas) the thickness often seen today in the Southwest.
The plethora of other grazers, living and extinct, reinforces
the impression.

Woodland, then, would have reached a non-savannah aspect
only in limited favorable spots on the flats and gentle
slopes and would have become dense (for woodland) only in
areas of rocky outcrops and shallow soils. As boreal forest
elements dropped out with decreasing elevation, woodland
trees would have been associated with sagebrush-grassland
and, eventually, with grassland only.

In part, this woodland model is based on the faunas. There
are few pure woodland creatures in the Southwest today. Armstrong (1972) listed nine Coloradan species primarily limited
to his saxicoline brush and/or pygmy conifer woodland communities that are close to our woodland classification.
These are Tamias dorsalis (cliff chipmunk), Perognathus fas-
ciatus (olive-backed pocket mouse), Peromyscus crinitus, P.
boylii (brush mouse), P. truei, Neotoma albigula, N. lepida,
Bassariscus astutus (ringtail), and Conepatus mesoleucus
(hog-nosed skunk). At least five of these regularly occur in
other habitats elsewhere. In much of our area, Urocyon tends
to occur in woodland, but also occurs lower and higher.

Peromyscus truei was tentatively identified in the Brown
Sand Wedge fauna and may well be represented among the many
reports of Peromyscus spp., but Tamias was rare or absent
from those sites within the woodland zone that did not also
indicate presence of coniferous forest, and Urocyon was rare.
Likely woodland did not exist essentially by itself until
fairly recent times. Indeed, some evidence suggests destruction of a richer understory than common today has occurred

within the past few centuries, probably as a result of overgrazing by sheep in post-Columbian time (Harris, 1977a).

The late Wisconsinan woodland, then, probably basically was
mostly sagebrush-steppe at higher elevations and a steppe-
savannah at lower, with the woodland elements present
throughout, but basic to few members of the mammalian fauna.

The grazing megafauna, common at virtually all sites, may
well have played a major role in maintaining not only the
woodland in this form, but also maintaining the coniferous
parklands hypothesized by various workers on the plains to
the east (and possibly in the northern portions of our area
if parkland rather than patchy distribution of boreal forest
was the case). Large herds of grazers (most of which, judging by living forms, also browsed to some extent) would have
tended to have opened up areas for colonization by trees by
local destruction of grasses and, at the same time, have
tended to limit tree growth by browsing, trampling, and use
as "back-scratchers."

Another approach to elucidating the makeup of the ecological zones was by counting the number of vertebrate taxa that
were represented in each vegetational zone. For example,
taxa occurring from the lowest sure woodland site (Rampart
Cave) to the highest site not containing evidence of sagebrush (Vulture Cave) were counted for Steppe-woodland. Taxa
from all sites within this span were counted even though some
sites showed no evidence of woodland. There are numerous
potential sources of error and the resulting figures must be
considered only as approximations. Based on the data established by the presence of marker species and midden plant
macrofossils, the predominant pattern should be one of generally increasing complexity from the lowest vegetative zone
up. That is, the lowest sites represented Steppe; Steppe

then was joined by woodland at higher elevations. In turn,
the Steppe-woodland was joined by sagebrush and this complex
(Sagebrush Steppe-woodlands) joined by boreal forest (Middle-
elevation Savannah) and becoming yet more complex with the
elevationally higher addition of tundra habitat (Northern
Highland Zone).

When such counts are made, the general pattern is one of
increase in number of taxa with increasing meters ee to the
Sagebrush Steppe-woodland Zone and then slight decreases to
the Northern Highlands Zone (approximately 38, 40, 111, 101,
and 96).

Sampling error for the fossil fauna can be alleviated
somewhat with the assumption that a taxon occurred in all
vegetative zones between its lowest site of occurrence and
its highest occurrence (although possibly not true in a few
cases, exceptions should be limited in number). Thus a taxon
not in faunas from the Sagebrush Steppe-woodlands would be
counted in that zone if it occurred in both the Steppe-
woodland below and the Middle-elevation Savannah zone above.
Under this assumption, the picture is, from lowest to highest
and with the number of sites in parentheses: 38 (2), 55 (4),
119 (8), 112 (18), and 96 (16). (Possibly the apparently
eastern forms †Dasypus bellus, †Glossotherium, Didelphis,
†Smilodon, and Procyon from the eastern plains sites should
be removed from the calculations, but this would not change
the picture appreciably).

Some of the pattern probably is the result of the distribution of the large fossil faunas, and certainly the Steppe
Zone is under-represented with only two sites. Nevertheless,
the pattern is strikingly different, even allowing for considerable error, from that of the present.

Accurate comparable figures are not easily available for

the current fauna, but an idea of present elevational distribution can be gained from data by Armstrong (1972) on the
distribution of the mammals of Colorado by Merriam's Life
Zones. The Upper Sonoran is approximately equivalent to our
grassland plus woodland; the Transition to our sagebrush; the
Canadian plus Hudsonian to our boreal; and the Alpine to our
tundra zone. From lower to higher life zone, the number of
species is 97, 63, 53, and 17. On the assumption that slope
differences and ecotonal sites might draw from adjacent life
zones, the taxa for each adjacent pair of zones was summed:
112, 71, and 53. Either way, the number drops drastically
with increasing elevation, unlike the Pleistocene stadial
condition.

Another pertinent way to look at present distribution in
terms of north-south and elevational differences is to note
the number of taxa represented in various states from south
to north. Findley et al. (1975) gave the number of species
for Arizona and New Mexico as 138 and 139 respectively, while
Colorado has 118 (Armstrong, 1972) and Wyoming 96 (Long,
1965). Again, there is a significant drop in the number of
taxa from south to north, the equivalent of increasing elevation ee. It should be noted that these four states each en-
compass elevational differences within their borders of more
than 3050 m.

Thus today, increase in either elevation or latitude results in a decrease in the number of mammalian taxa, a vast-
ly different pattern than seen in the Wisconsinan-age faunas.

The primary reason for the difference between the Wisconsinan pattern and that of today appears to lie in the relative complexity of the various zones. The Northern Highland
Zone actually included a fauna that today tends to be sub-
divided to considerable degree into four major vegetative

habitats: tundra, boreal, sagebrush, and grassland; the upper portion of the Middle-elevation Savannah Zone into three
groups and the lower part of this zone into four groups; the
Sagebrush Steppe-woodland Zone included three groups.
Steppe-woodland, however, encompassed only two vegetative
types and the Steppe Zone only one.

The reason for this pattern appears to be in large part
that boreal mammals and plants tended to extend to much lower
elevations than today, while the upper boundary of present-
day lowland forms retreated downslope to a much lesser degree. Consequently, the upper zones telescoped into one
another--but this does not appear to be a narrowing of the
lower zones, as suggested by Wright et al. (1973) and Brackenridge (1978), but an overlapping, with the lower borders of
all zones extending far below their present elevations (the
evidence strongly suggests that the taxa reacted individual-
istically to environmental factors, but, with notable exceptions, the results were very roughly similar to unitary zone
shifts). In part, this shift was due to the increase in
effective precipitation allowing a spread of taxa now limited
by aridity. However, as discussed by Harris (1970b), much of
this effect, particularly in the south, may have been a combination of cooler air temperatures combined with mean insolation values approaching those of today. Although cloud
cover may have increased to some degree, insolation during
clear weather would have been little changed from that of today at any given site. Slopes protected from insolation
would have been subjected to the full effects of the cooler
air temperatures, while slopes receiving insolation would
have intercepted a greater amount of energy than do slopes
that today lie in regions of equally cool air masses. The

result would be an exaggeration of slope differences over those of the present.

Flatland sites, however, should have differed climatically from those of today almost entirely by the effect of differences from the present in air mass temperatures (since little slope effect would have been present) plus the effect of any differences in effective precipitation. Thus any increase in biotic complexity should be attributable to air mass temperature and precipitation changes alone, whereas areas with moderate to high relief should show proportionately greater change in complexity over that of today because of the additional exaggeration effect. Unfortunately, flatland sites usually are open sites with few preserved (or recovered) microfaunal taxa, making it dificult to test this hypothesis.

The overall effect in areas of high relief was major movement down protected slopes by the lower borders of forms that today are restricted by warm temperature and/or aridity, but only moderate retreat of upper borders down exposed slopes by forms limited now by cool temperatures and/or excessive effective precipitation.

A very broad idea of Wisconsinan changes in zones can be gained by comparing their modern lower boundaries with those of the Pleistocene. Using data from Lowe (1964) and Armstrong (1972), approximate lower boundaries were translated into meters ee. The present lower border of the alpine zone is about 7600 m ee; the Wisconsinan lower boundary was about 6100 m ee, a difference of ca. 1500 m. Depression of boreal forest elements was similar, from a present 6550 m ee to about 5025 m ee, a difference of 1525 m.

In the absence of a recognizable Transition Life Zone forest in the fossil record, sagebrush is used as a Transition zone marker. With a lower boundary to the present zone of

ca. 6000 m ee, and a very approximate Wisconsinan lower border of the Sagebrush Steppe-woodland Zone at about 4550 m ee,
the difference is 1450 m. Woodrat midden evidence indicates
occurrence to a considerably lower elevation. Van Devender
and Spaulding (1979) recorded Artemisia tridentata at 4399 m
ee (Ajo Mountains, AZ), 4363 m ee (Artillery Mountains, AZ),
and 4133 m ee (Picacho Mountains, AZ). The latter site is
lower than any of our vertebrate sites and might indicate our
lower sites still were within the Sagebrush Steppe-woodland
Zone. These lower sites may have represented scattered populations too scanty to have supported a sagebrush fauna or
perhaps isolated from continuous bodies of sagebrush. The
Artillery Mountains site also contained creosotebush, indicating a somewhat marginal situation. At present, the sagebrush zone is left interpreted here as given earlier, but the
1450-m depression certainly should be considered minimal and
more investigation of the Wisconsinan sagebrush zone lower
limit is needed.

Woodland, as defined faunally, showed a depression of ca.
1050 m (present woodland, 5425 m ee; Wisconsinan woodland,
4375 m ee). Once again, midden evidence indicates a full-
glacial depression below that recognized faunistically. Van
Devender and Spaulding (1979) recorded pinyon pine as low as
4133 m ee (Picacho Mountains); several other sites lie below
4200 m ee (Whipple Mountains, CA; Tucson Mountains, AZ; New
Water Mountains, AZ). Although the two lowest fossil sites
considered here seem to have lacked woodland, presence of
such is a distinct possibility (particularly for Jimenez
Cave), and a pure grassland habitat may have been entirely
absent from our record.

Thus, although the woodland zone was depressed somewhat
less than the upper zones based on somewhat unsatisfactory

faunal evidence, midden data indicate depression probably was
similar.

The lower boundary of the Late Pleistocene Upper Sonoran
Life Zone cannot be measured directly by these data; absence
of any Lower Sonoran site indicates the Upper Sonoran/Lower
Sonoran boundary was below any of our sites. Lowe (1964)
placed the present boundary at the equivalent of about 4500 m
ee. If we assume a depression of 1000 to 1500 m, the Wisconsinan boundary would have been between 3500 and 3000 m ee.
The former figure is close to the lowest elevation ee in our
region. Midden sites in the Chihuahuan Desert record pinyon
to as low as 3757 m ee (Van Devender and Spaulding, 1979) and
in the west (Chemehuevi Mountains, CA) to 3950 m ee (Wells,
1979). Wells also reported juniper woodlands from subtropical Sonora, Mexico, in the southern lowlands of the Sonoran
Desert. It appears the lower boundary of the Upper Sonoran
Life Zone (sensu lato) was lowered to the same general degree
as were the lower boundaries of the higher zones, or even to
greater extent.

In general, depressions of most vegetational life zones as
based on the faunas tend to be somewhat greater than the published figures based on midden evidence. This may in part
reflect our use of a general figure for present lower boundaries and in part the different environmental sampling
represented by middens compared to non-midden sites (particularly cave sites). Middens sample mostly from the immediate
vicinity of the site (within ca. 100 m), whereas many fossil
sites contain microfaunal elements carried in by predators
from a larger region, though probably mostly from within a
few kilometers and virtually all from less than 32 km (Harris, 1977b).

Upper boundaries of the Wisconsinan-age vegetative types

seem to have been relatively little changed from those of the present. A very rough estimate of the current upper boundary of the boreal zone (top of the Hudsonian Life Zone) is ca. 7500 m ee, while our highest site at 7017 m ee showed evidence of boreal elements. Sagebrush reaches the equivalent of ca. 7650 m ee in Wyoming today (Cary, 1917); again, the upper boundary of the Late Pleistocene sagebrush zone must have been above 7017 m ee. The current upper boundary of the Upper Sonoran Life Zone may be estimated very roughly as 6100 m ee (several estimates in the literature vary from ca. 5800 to 6450 m ee); the upper boundary of the Pleistocene woodland as a significant vegetative type is estimated to have been slightly over 5400 m ee.

Thus the Wisconsinan lowering of the upper boundaries of all of the zones for which we have evidence probably was in the general range of 600 to 700 m.

The Sagebrush Steppe-woodland Zone seemingly marked an important climatic-ecologic boundary. Thirty-seven taxa did not make the transition from the Northern Highland Zone to the Middle-elevation Savannah Zone; 35 taxa failed the Middle-elevation Savannah/Sagebrush Steppe-woodland transition; but about twice as many taxa (69) appeared in the Sagebrush Steppe-woodland than in the zone below (27 failed the Steppe-woodland/Steppe transition, but likely the limited Steppe Zone sample is biasing this figure). Comparable figures based on assumption of occurrence at all sites between lowest and highest sites of appearance are 26, 29, 60, and 18).

The implication of these figures is that ecologic parameters changed rather gradually to the bottom of the Sagebrush Steppe-woodland Zone, but there was then a relatively abrupt biologically important change. This may in part be due to

the distribution of sites. All four Steppe-woodland sites
used here are in southern California, southern Nevada, or
adjacent Arizona, places strongly affected now by the rain
shadow of the high Californian mountain ranges and, according
to Wells (1979), more arid during the late Wisconsinan than
woodland farther to the east. Increasing aridity, probably
caused in part, at least, by increasing temperatures in the
high-insolation areas of the Southwest, seems most likely to
have been an important factor here. Many of the taxa that
dropped out were those requiring moderate to heavy cover,
particularly of grasses, while few such continued into the
Steppe-woodland or Steppe zones. The appearance is of having
moved from a sagebrush-grassland-woodland region with relatively heavy grass stands to a zone where grassy ground cover
was too scanty to have regularly supported most microtines,
shrews, and the like. More and better distributed low-
elevation sites are needed to test the validity of this.

FULL GLACIAL--WEST COAST DIVISION

The West Coast Division includes a series of sites lying west
of the crest of the Cascade-Sierra Nevada ranges. Ten sites
are considered, but ages are uncertain for many and some may
be older than latest Pleistocene. Each site is considered
briefly from highest elevation ee to lowest (apparent absence
of published elevations for several sites results in some
estimations).

Sites

Potter Creek Cave, Shasta Co., CA.   457 m, 4844 m ee (Hutchison, 1967).

This site lies within the mountainous terrain of the Transition Life Zone now. Sinclair (1904) believed the fauna to be older than late Wisconsinan.

Kellogg (1912) noted that the rodent and lagomorph fauna included members now typical of Upper Sonoran, Transition, and Canadian life zones and that an assumption of a more mesic, partially forested situation might allow the Canadian forms into the area without an assumption of a much colder climate. Thus a situation analogous to that of the Interior may be represented, with greater effective moisture (partly, perhaps, a result of cooler conditions) allowing movement downslope of higher-elevation forms without concomitant descent of upper boundaries of lower zones.

Samwel Cave, Shasta Co., CA.   Ca. 455 m, ca. 4842 m ee (Kellogg, 1912; Stock, 1918).

Samwel Cave, close to Potter Creek Cave, is in similar type country and under similar present vegetative conditions.

The fauna, presumed to be late Wisconsinan, was generally similar to that of Potter Creek Cave and indicated a similar environment.

Hawver Cave, Eldorado Co., CA.   393 m, 4566 m ee (Stock, 1918).

This site is in the Sierra Nevadan foothills, in an area of predominantly Upper Sonoran vegetation and fauna.

The fossil fauna indicated by the presence of sewellel (Aplodontia rufa) that humid forest was nearby. Open grassland also seems to have been present. Again, then, somewhat more mesic conditions than occur today probably was the case.

McKittrick, Kern Co., CA. 320 m, 4119 m ee (Schultz, 1938).

McKittrick is an asphalt-seep site similar to Rancho La Brea. Its location is in the southwestern foothills of the relatively arid Central Valley of California. To the west, the Temblor Range rises, supporting sparse brush and occasional small trees.

The extant members of the McKittrick fauna indicated little change from conditions of today other than the presence of a standing body of water in the area. Floral evidence indicated _Juniperus utahensis_ was somewhat out of place compared to today.

Maricopa, Kern Co., CA. 260 m, 4005 m ee (J. R. McDonald, 1967).

This site, although associated with asphalt, probably was a waterhole. It is located in the San Joaquin Valley about 50 km east of McKittrick. A date of 13,860 BP is available (Nowak, 1979).

Kurten and Anderson (1980) suggested the climate probably was cooler and wetter, but the few published identifications do not seem to allow meaningful interpretations.

Channel Islands, Santa Barbara Co., CA. Ca. 3790 m ee (Kurten and Anderson, 1980).

Several of the islands are grouped together here. The elevation is an approximation, since sites vary in elevation through 100 m or more. ^{14}C dates indicate late Wisconsinan deposits.

Little environmental information is supplied by the fauna. Dwarf mammoth, mastodon, and two extinct species of <u>Peromyscus</u> are of interest because of the insular nature of the fauna.

Carpinteria, Santa Barbara Co., CA. Ca. 3696 m ee (Kurten and Anderson, 1980; Wilson, 1933).

This site, near the coast between Ventura and Santa Barbara, is believed to have been deposited under cooler, moister conditions during the late Wisconsinan; the site may be older than this, however, with Warter (1976) citing a date of >38,000 BP.

Tree squirrels and shrews (<u>Sorex</u> <u>trowbridgi</u>) imply at least moderately thick forests were in the vicinity, <u>Sylvilagus</u> <u>bachmani</u> suggests that there was chaparral near the site, and a number of forms indicate grassland or open grassy woodland was about.

Floral remains imply conditions were closely similar to those now found on the Monterrey Peninsula, some 330 km to the north. Presence of redwood remains (Warter, 1976) probably showed occurrence in the nearby highlands, far south of its present range.

Rancho La Brea, Los Angeles Co., CA. Ca. 3653 m ee (Stock, 1956).

This site is surrounded by the city of Los Angeles. It lies on the Los Angeles Plain some 5 km from the southern front of the Santa Monica Mountains. Dates on asphalt pit deposits ranged from Holocene to ca. 40,000 BP.

The fauna was similar in its extant members to the present day fauna. Warter (1976), working with plant material from Pit 91 (25,000 to 40,000 BP), recognized a close-cone pine

forest at the site itself, with plants similar to those on
the Monterrey Peninsula. She interpreted a chaparral-
foothills woodland and a coastal redwood element as having
been transported from the nearby highlands.

La Mirada, Los Angeles and Orange counties, CA. Ca. 3653 m
ee (W. Miller, 1971).

   This site is close to both floodplain and hill habitats.
Miller (1971) suggested the nearby hills probably supported
woods and the Los Angeles Plains were covered by grasses.

Discussion

Interpretation from a faunal viewpoint is disappointing.
Sites south to Carpinteria generally do give faunal indications of having supported a slightly more complex community,
including elements that today occur under higher, moister
conditions, but resolution is poor. Most information on past
conditions of the southern sites came from floral analyses.

   Several factors may be involved, including the generally
equable nature of the Pacific Coast climate, the distribution
of the sites, and my unfamiliarity with many of the taxa involved. Likely, Pleistocene changes actually were less dramatic in this area than inland--many of the inland changes
appear to have been a result of increased equability, of decreasing summer temperatures (and consequent increased effective moisture as well as probable actually increasing precipitation), and of winter temperatures lacking the extremes of
today. The West Coast generally already has these features,
and the major change probably was more of a relatively simple
shift southward and downslope than in the Inland Division.

THE SANGAMONIAN

There are notable problems in considering the Sangamonian. Since this interglacial lies beyond the range of radiocarbon dating and other direct dating methods seldom can be applied, assignment to the Sangamonian usually is based on stratigraphic and biologic evidence. This adds a considerable degree of circularity, where frequently we are in the position of recognizing the interglacial deposits by their faunas and then characterizing the Sangamonian fauna on the basis of it occurring in interglacial deposits.

This might not be a great problem if we had two clear-cut faunas, glacial and interglacial. We do not. Faunas of the mid-Wisconsinan are intermediate to greater or lesser degree, based on our limited evidence. But we do not know how different the most extreme interstades were from the interglacial conditions (nor, for that matter, how different some mid-Wisconsinan stades were from such conditions). Consequently, we tend to assign a non-glacial fauna to an interstade if the ^{14}C date is finite and to an interglacial if beyond the ^{14}C dating method.

Stratigraphic position merely tells us that one deposit is older or younger than another. A combination of faunal and stratigraphic data may help if the faunal material includes forms that have had evolutionary or migrational events that are securely tied to a time frame. Thus the presence of †Panthera leo atrox in the American Falls fauna is taken to indicate Sangamonian age (Kurten and Anderson, 1980) because that cat is not known from earlier deposits; however, the same fauna earlier was presumed to be Illinoian because

†Bison latifrons was unknown from post-Illinoian deposits until relatively recently.

The sites included here probably are all post-Illinoian; some, however, may well be mid-Wisconsinan (in a broad sense) rather than Sangamonian.

Faunas tentatively treated here as Sangamonian are: American Falls, Fort Qu'Appelle, Medicine Hat Fauna 7, Mesa De Maya, Newport Bay Mesa Loc. 1066, San Pedro, Saskatoon, and Silver Creek.

Some areas, such as Medicine Hat, Fort Qu'Appelle, and Saskatoon, were under ice during the full glacial and quite possibly during Wisconsinan interstades. Thus the presence of faunas at these sites at least denote ice-free local conditions. Harington (1978) indicated the Fort Qu'Appelle and Saskatoon faunas may be interstadial. Unfortunately, most of the faunal elements from these northern sites were large, now-extinct herbivores, such as †Bison latifrons, †Equus spp., †Camelops hesternus, †Mammuthus columbi, †Hemiauchenia sp., †Cervalces roosevelti, and †Symbos cavifrons; or wide-ranging carnivorous types such as †Canis dirus. Apparently open grazing lands were well represented, while †Cervalces and †Symbos may indicate browsing resources.

Some other animals, primarily from Medicine Hat fauna 7, gave additional data. Cynomys cf. ludovicianus does not reach the area today, and this grassland form presumably represented somewhat warmer and possibly drier conditions than at present. Most other forms from Medicine Hat also indicate open grassland was present, though perhaps approaching some aspect of tundra (Rangifer tarandus); whether this apparent discrepancy between warmer grasslands and some type of tundra-grassland is due to climatic variation during deposition or to one or another taxon occurring under conditions different than inhabited today is unclear. A few forms indicate that probably thickets of trees or perhaps riparian forests were in the area. These included <u>Lynx</u> <u>canadensis</u>, ? <u>Alces</u>, <u>Cervus</u> <u>elaphus</u>, <u>Odocoileus</u>, <u>Erethizon</u>, and possibly †<u>Megalonyx</u>. The Canadian sites probably were mostly open grasslands or parklands with forested conditions along the river valleys.

The remaining sites suggest Sangamonian conditions (if that is what is represented) were similar to conditions of today. At Silver Creek, Utah, <u>Brachylagus</u> was some 80 to 95 km east of its present range (Miller, 1976) and at a high elevation for that taxon. However, the overall fauna required only extinction of several large herbivores and carnivores plus the presence of a marsh to otherwise closely approach the situation of today.

The American Falls site (5941 m ee) again was quite similar in its extant fauna to that of today, though possibly it required slightly more effective precipitation.

The Mesa De Maya fauna (5759 m ee) was interpreted by Hager (1974) on the basis of mammals, molluscs, and pollen to have existed under more equable temperatures than those of today, and with cooler summers possibly accounting for slightly more effective moisture. These conditions were much the same as interpreted for the Dry Cave interstadial deposits, and possibly an early Wisconsinan or late Illinoian interstade was represented rather than full Sangamonian conditions. The presence of two extinct mice (†<u>Peromyscus</u> <u>cragini</u> and †<u>P</u>. <u>progressus</u>) indicates a fairly old deposit.

The two southern Californian, lowland faunas (San Pedro and Newport Bay Mesa Loc. 1066) showed few differences from modern conditions.

The meager evidence, if accepted uncritically, suggests

warmer conditions than today occurred in the far north, more equable conditions to conditions similar to those of today in the Inland Division, and conditions close to those of today on the West Coast.

THE INTERSTADIALS

Many of the same problems in recognizing Sangamonian sites plague recognition of interstadial faunas, as discussed earlier. Thus, as used here, "interstadial" refers essentially to sites falling between the early Wisconsinan and late Wisconsinan full-glacial times.

The Dry Cave interstadial sites are discussed at greater length in the following section, but basically indicate climatic conditions were more similar to modern conditions than to full-glacial, but probably were somewhat cooler, more equable, and with somewhat greater effective precipitation.

Two of the Medicine Hat faunas in Alberta were deposited under non-glacial conditions after Sangamonian times: Fauna 4, late mid-Wisconsinan; Fauna 5, early Wisconsinan (^{14}C dates on wood 1 to 2 m above the fauna dated at 37,900 ± 110, GSC-1442; 38,700 ± 1100, GSC-1442-2) (Harington, 1978).

†Nothrotheriops was present in Medicine Hat Fauna 4, far north of the late-glacial records in the Interior Division and at somewhat higher elevation ee. Harington (1978) interpreted the habitat as a grassland with patches of trees and shrubs and with moist areas.

Medicine Hat Fauna 5 had more small mammals represented, including Cynomys cf. ludovicianus, Spermophilus richardsoni, and Microtus. The Cynomys was north of its present range.

The microfauna and megafauna alike suggest that the environment was that of a prairie grassland (Harington, 1978).

In Oregon, Fossil Lake appears to fall into the mid-
Wisconsinan time span. A ^{14}C date on snail shells from
"several feet" above the fossil mammal zone and 4.3 m below
the top of the lake beds was 29,000 BP (Allison, 1966). Many
or most of the bird bones came from above the dated level,
but Allison believed the mammals may have dated from considerably earlier than 29,000 BP since they occurred at or near
a major disconformity. He suggested a mid-Wisconsinan or
possibly an early Wisconsinan age.

The area was described by Allison (1966). The climate is
dry (199 mm per year at a station 22.5 km southwest of the
site). The area formerly was occupied by pluvial Fort Rock
Lake, with a depth of more than 60 m. Present vegetation
includes sparse growth of sagebrush and other arid shrubs,
bunch grass, and salt grass; scattered junipers occur to the
northwest on higher ground and there is a stand of ponderosa
pine growing on sand on the basin rim to the east. A small
playa lake holds water only during wet weather, but reported-
ly was permanent early in the 20th century.

The few clues given by the fauna suggest, aside from the
extensive aquatic habitat, that there was a well developed
grassland. Several forms appeared here and elsewhere that
seem to have been essentially absent from the Interior Division during the late glacial. These included †Glossotherium,
Panthera onca, and two species of †Platygonus. Two species
of small mammals, †Thomomys vetus and †T. scudderi, are extinct (Elfman, 1931, and Kurten and Anderson, 1980, considered these synonyms of T. townsendi, but Russell, 1968,
did not).

Tule Springs Unit B-2 (Mawbry, 1967; Haynes, 1967) is old-

er than 40,000 years. Pollen indicated the valley around the
site was dominated by sagebrush, with ponderosa pine and
sagebrush on the nearby bajadas; Abies must have inhabited
the upper bajadas (or have occurred even lower) (Mehringer,
1965). How much of this deposit represents interstadial conditions is unsure, and the deposits may even represent stadial conditions. The fauna is of relatively little interpre-
tive help, with Microtus sp. the only small mammal present of
any diagnostic value. The normal Pleistocene large herbivores occurred. Thus, although grasses were present, little
else is clear.

Rainbow Beach, in southern Idaho, seems to fit into the
pre-full glacial period, with dates of 21,500 ± 700 (WSU-
1423) and 31,300 ± 2300 (WSU-1424) on bone collagen. The
fossils were in deltaic sediments deposited by spillover from
pluvial Lake Bonneville.

Centrocercus urophasianus and Brachylagus idahoensis indicated the presence of sagebrush, while †Glossotherium harlani, Cynomys cf. leucurus, Spermophilus richardsoni, and
Microtus cf. montanus are forms expected to have been associated with grasses. The Dam local fauna, dated at 26,500 ±
3500 BP on bone reflects similar conditions to those at Rainbow Beach (Kurten and Anderson, 1980).

In general, these interstadial sites showed much the same
kinds of differences from modern conditions as did those at
Dry Cave.

THE DRY CAVE SITES--CHANGE THROUGH TIME

Sites

A series of sites at one locality, but differing in times of
deposition, allows tracing of changes through time while
standardizing elevational and topographic relationships.
Such a series was found in Dry Cave, in southeastern New
Mexico.

The cave is located almost midway between the Guadalupe
Mountains to the west and the Pecos River to the east, lying
some 24 km west of Carlsbad. The topography is of rolling
limestone hills moderately dissected by drainageways. Elevation is 1280 m (4758 m ee) at the cave entrance. Both present and past entrances into the cave system are on upper
slopes, so that inwash is limited in origin to a few hec-
tares.

Present vegetation is ecotonal between Upper and Lower
Sonoran life zones. Such Chihuahuan Desert forms as lechu-
guilla (Agave lecheguilla), all-thorn (Koeberlinea spinosa),
and a few creosotebushes (Larrea divaricata) mingle with low
junipers (Juniperus erythrocarpa).

Much of the present surface is of bare limestone, with
soil being mostly limited to crevices and topographically low
areas. Probably soil loss has been historic, at least in
part, due to overgrazing by sheep.

The sites, excavated by the University of Texas at El Paso
(UTEP), have been studied to varying degrees. Some details
have been published (Applegarth, 1979; Harris, 1970, 1977b,
1980; Harris and Crews, 1983; Harris and Mundel, 1974; Harris

and Porter, 1980; Harris et al., 1973; Holman, 1970; Magish and Harris, 1976; Metcalf, 1970).

The cave itself is an extensive maze system on several levels. Fissures have intersected the surface in different areas at various times. Only one of these, the Entrance Fissure, currently allows access. The organic materials have entered the system through these fissures by inwash, transportation by mammalian and avian predators, and by utilization of the system for dens, nesting places, and the like.

Two sites mentioned in the literature (Harris, 1970) have proven to be of little value. Locality 3 consisted of a series of passageways with small amounts of material on or near the surface. It almost certainly had a collection built up over a considerable span of time and probably included Holocene material as well as Wisconsinan. Locality 12 was a mixed deposit, receiving Pleistocene material from one or more sources (including TT II), but also receiving modern material from the present-day entrance.

A summary of the sites, their dates, and interpreted climatic conditions is given in Table 6; each, arranged from oldest to most recent, is described in more detail below.

Room of the Vanishing Floor (UTEP Loc. 26, 27).   33,590 ± 1550 (TX-1773).

The ^{14}C date was on bone carbonates. Although such dates are untrustworthy, a pre-full glacial time is indicated. In previous publications, this and the following two sites have been referred to as interstadial. As mentioned earlier, however, exact placements on the stadial-interstadial continuum is impossible at present. For ease of presentation, however, they'll continue to be referred to here as interstadial.

Two separate bone-quarrying sites (originally designated

UTEP 26 and UTEP 27) are represented from different sides of the fill of a single chimney; bone scrap from both localities was combined to get a sufficient sample for dating. The fossils entered the chamber from a now-closed fissure (chimney). All surface traces of the original openings of this and the other interstadial sites have been obliterated.

Lost Valley (UTEP Loc. 1, 17). 29,290 ± 1060 (TX-1774).

This site also was dated on bone carbonates. The source of the fauna is a now-clogged large fissure that intersects the ceiling of a large chamber. Material from the fissure landed upon a sizable ledge beneath (UTEP 1), with much of the material cascading into a large pit originally designated as UTEP 17. The size of the fissure makes it likely that its walls served as a bird roost.

Sabertooth Camel Maze (UTEP Loc. 5). 25,160 ± 1730 (TX-1775).

A passageway (UTEP Loc. 2) opening onto the side of the Lost Valley Pit has accumulated bones from an unknown source. This opening slopes downward as a wide, low-ceilinged maze to another area of accumulation (UTEP 5). The material has been lumped for interpretation and referred to as Loc. 5. The ^{14}C date was on bone carbonates.

Pit N & W of Animal Fair (UTEP Loc. 122). Est. >20,000 BP.

The three previous sites are located far from the present open entrance to the cave system. UTEP 122 is located within the fissure that forms the present entrance, but in the now-filled portion north of the present opening. As the fissure filled from north to south, the opening into the system migrated southward; thus deposits at the same depth should differ in age, those to the north being older.

TABLE 6. Dry Cave sites arranged in order of presumed age, from oldest to youngest, with a rough indication of inferred climatic conditions for each.

Site	Date	Climate
Rm. Vanishing Floor (UTEP 26, 27)	33,590	warm summer, cool winter
Lost Valley (UTEP 1, 17)	29,290	warm summer, cool winter
Sabertooth Camel Maze (UTEP 5)	25,160	warm summer, cool winter
Pit N & W Animal Fair (UTEP 122)	----	slight summer cooling
Animal Fair, older (UTEP 22)	----	cooler, moister
Camel Room (UTEP 25)	----	cool summer, cold winter
Animal Fair, younger (UTEP 22)	15,030	cool summer, cold winter
Harris' Pocket (UTEP 6)	14,470	cool summer, cold winter
Human Corridor (UTEP 31)	----	warmer
Bison Chamber (UTEP 4)	----	as above
Stalag 17 (UTEP 23)	11,880	as above
TT II (UTEP 54)	10,730	warmer, drier
Entrance Chamber (UTEP 24)	<11,880	as above to modern

This fossil-producing chimney is to the north and at a
somewhat higher level than the elongate site known as Animal
Fair (Loc. 22, see below), and Loc. 122 should be older than
15,030 BP by an appreciable degree.

Camel Room (UTEP Loc. 25).   Est. >15,000, <20,000 BP.

By virtue of its position (north of Animal Fair and the
present entrance, but nearer the surface of the filled fissure than is UTEP 122) and fauna, Camel Room appears younger
than Loc. 122 and younger than the older deposits in Animal
Fair, but older than the single date from Animal Fair.

Animal Fair (UTEP Loc. 22).   15,030 ± 210 (I-6201).

The ^{14}C date was from the southern end of the fissure itself, at about the same level as the upper surface of the
fill in the adjacent chamber (Animal Fair proper). The site
included material from within the fissure (informally, Charlies Parlor) and deposits that spilled from the fissure into
chambers on either side (Hampton Court to the west and Animal
Fair to the east of the fissure). At about the dated time,
further access into the adjacent chambers was blocked by
clogging, but deposition continued in Charlies Parlor.

Because of the north to south retreat of the entrance to
the cave, the fissure and chamber fill should be increasingly
ancient with distance from the southern end, as well as the
deeper deposits being older than the upper. This is borne
out by the herptile material studied by Applegarth (1979) and
by the present findings. Material assigned to this site and
stratigraphically higher than the dated level could range
from as little as a few hundred years after 15,030 BP to as
much as 1000 to 1500 years. Material from Animal Fair and
Hampton Court, depending on depth and distance north of the

dated material, could be many thousands of years older.
Since study of the site is still in a relatively early stage,
no differentiation is made in the listed fauna according to
relative age, but pertinent differences are mentioned in the
discussion and interpretation of the Dry Cave faunas.

Harris' Pocket (UTEP Loc. 6).   14,470 ± 250 (I-3365).

Material from this site, ^{14}C dated on rodent feces, probably was redeposited by water from near the base of another
fissure, Bison Fissure. Although blocked, this fissure is
clearly visible on the surface as a clogged sink. The suspected pathway of deposition of UTEP 6 is too constricted to
trace physically.

Human Corridor (UTEP Loc. 31).   <15,000, >12,000 BP.

This site intersects Entrance Fissure above the southern
end of Animal Fair; some material recognizable as representing the same individual has been recovered from the higher
fissure deposits of Charlies Parlor (Animal Fair) and from
this site.

Bison Chamber (UTEP Loc. 4).   Est. <14,470, >10,730 BP.

Deposits in this chamber came from Bison Fissure and are
believed to have been laid down after blockage of small
passageways feeding Loc. 6, but earlier than deposition at TT
II nearer Bison Fissure.

Stalag 17 (UTEP Loc. 23).   11,880 ± 250 (I-5987).

This chamber has been connected with the present entrance
area by a test pit. A subtle change in color and texture of
the fill at about 3.8 m below the surface of the Entrance
Chamber fill is believed to mark the end of the Pleistocene

deposition and the commencement of the Holocene. Charcoal flecks gathered from within about a 20-cm span below this boundary supplied the material for the ^{14}C date.

TT II (UTEP Loc. 54). 10,730 ± 150 (I-6200).

This site lies on a ledge at the base of a portion of Bison Fissure. The date was on bone collagen and, as such, may be somewhat too young. Material from this site has overflowed the ledge edge into a pit (Balcony Room, UTEP 12), where the material has mixed with Holocene matter entering from the present entrance.

Entrance Chamber (UTEP Loc. 24). <11,880 to present.

Fill continuous with that of Loc. 23 was deposited during the Holocene. A ^{14}C date of 3135 ± 165 (I-6199) on charcoal comes from the 80-100 cm level.

Only the woodrat fauna has been studied in detail.

Discussion.

The sites fall into two major groupings according to their faunas and ages. The Sabertooth Camel Maze, Lost Valley, and the Room of the Vanishing Floor faunas differ in several important ways from those of younger age. The Pit N & W of Animal Fair (UTEP 122) may be transitional, but appears to be closer to the older triad. The older deposits of Animal Fair seem to be younger than those of Loc. 122, but by an unknown amount.

Appearing in the older group exclusively or in that group and Loc. 122 are (among others) Lepus californicus, Spermophilus variegatus, Cynomys ludovicianus, ? Tamias, Panthera

onca, and †Tapirus sp. An undescribed genus of rabbit (listed in the appendices as "†Leporidae, extinct rabbit," and
currently under study by Brett Russell and myself) also
occurred in these sites, as did the extinct †Corvus neomexi-
canus (Magish and Harris, 1976) and two undescribed species
of woodrats (Harris, in press,a,b). None of the extinct
forms appeared in the relatively small sample from Loc. 122.

Notably absent from the older sites, but present in the
younger group of sites, were Lepus townsendi, Spermophilus
cf. tridecemlineatus, Thomomys talpoides, Sorex monticolus,
Sorex cf. nanus, Sorex merriami, Cryptotis parva, Thomomys
bottae, Neotoma cinerea (present in Loc. 122), Neotoma ?
goldmani, Microtus longicaudus, Microtus mexicanus (present
in Loc. 122), Lagurus curtatus, and Cynomys (Leucocrossuromys).

The older deposits of Loc. 22, in common with those of the
early group of sites through Loc. 122, included Sylvilagus
auduboni (or possibly S. floridanus); S. nuttalli occurs
throughout the Dry Cave sequence except for the Holocene
Entrance Chamber.

Although likely sampling error accounts for some differences, it almost certainly does not for many. For example,
sample size for Sylvilagus from the late sites is large, but
no specimen can be identified as S. auduboni. Likewise, L.
californicus has been looked for within the younger sites; it
has been identified from the older end of the Entrance Fissure deposits (Loc. 122), but from no site known to be less
than about 15,000 radiocarbon years old. Special effort has
been taken to check for L. townsendi from the older deposits,
without success.

Presence of such taxa as S. auduboni, L. californicus, and
Sigmodon give the older sites a somewhat modern look, but

other taxa indicate that modern conditons were not present.
Several of the taxa are extinct, and such modern forms as
Sylvilagus nuttalli and Pitymys ochrogaster do not approach
the area today, though appearing in the full-glacial sites
and presumably indicative either of more effective moisture
than present today or of a different seasonal distribution of
precipitation.

Other taxa supply climatic data, also. Localities 5 and
26 both had desert tortoise (Gopherus agassizi) represented
(Van Devender et al., 1976) and Loc. 5 had an extinct tortoise, †Geochelone wilsoni (Moodie and Van Devender, 1979).
These tortoises are widely considered as indicating relatively mild winters, and the eastern limit of G. agassizi currently "coincides with the eastern limits of several, frost-
sensitive Sonoran Desert dominant plants" (Moodie and Van
Devender, 1979).

The occurrence of the modern California roadrunner (Geococcyx californianus) from Loc. 26 was used by Harris and
Crews (1983) as indicative of warmer summer temperatures than
occurred during the full-glacial, when †G. conklingi (interpreted as †G. californianus conklingi by Harris and Crews)
inhabited the Southwest. This is supported by presence of
Sigmodon, which probably required a July mean temperature of
24°C or above (Mohlhenrich, 1961).

Overall, the fauna indicates there were mild winters and
warm to hot summers, more effective moisture, and fair grasslands. Presence of Sylvilagus nuttalli suggests at least
local stands of woodland trees were present, as does the very
tentatively identified chipmunk (? Tamias) and the occurrence
of Urocyon at all three sites (the gray fox was rare in the
Interior Division full-glacial sites).

Locality 122 fits into the general pattern, although the

addition of Neotoma cinerea might indicate slight cooling
beyond that seen in the earlier interstadial sites.

Within this suite of older sites, few differences between
faunas not attributable to sampling error are seen, and the
tentative conclusion is that similar climatic regimes were
represented by these sites.

Camel Room (Loc. 25) had a relatively small fauna, but
Lepus californicus had been replaced by L. townsendi and Sylvilagus auduboni was absent. †Hemiauchenia appeared for the
first time, but its earlier absence probably was the result
of sampling error. Urocyon appeared for the last time during
the Dry Cave Pleistocene. Neotoma cinerea now was the common
woodrat. Assuming correct stratigraphic assignment, the
Camel Room deposits mark the first entirely stadial deposits
at Dry Cave.

The Animal Fair deposits, however, seem to record the
actual transition into full-glacial times. Although analysis
is in its early stages, it is clear that Pitymys is common in
the older levels, becoming very rare or absent in upper deposits near the southern end of the site. Lagurus is present
throughout, but Microtus appears relatively limited in occurrence in the older deposits (and probably consisted mostly or
entirely of M. mexicanus), becoming frequent only later.
Cryptotis parva, Sylvilagus auduboni (or S. floridanus), and
Perognathus (Perognathus) occur rarely in the older levels
and apparently are absent in the more recent. Specimens of
Neotoma ? goldmani are known only from the earlier deposits.
Microtus pennsylvanicus and Sorex cf. nanus thus far are
known by single occurrences in the late deposits. The impression at this point is a transition from a relatively
cool, dry-aspect sagebrush grassland (with woodland?) to a
more mesic sagebrush-steppe woodland. Nevertheless, the

earlier deposits denote notably cooler, moister conditions than those of the interstadial sites.

Except for the Entrance Chamber site, the remaining Dry Cave sites were deposited under stadial conditions. There is indication of some climatic progression, however. The earlier stadial sites show that conditions then were cool and moist. Absence of the large tortoises throughout suggests winter temperatures were too cold for their survival. However, Gopherus occurred in late stadial times a short distance to the west (Shelter Cave, in south-central New Mexico) at about the same latitude and even slightly greater elevation. Unless the edge of the plains country at Dry Cave had a far different climate (and the faunas overall suggest otherwise), the Dry Cave area probably lacked extremes of winter cold and conditions were only slightly beyond desert tortoise parameters. That ample cover and cool summer temperatures were present is testified to by the occurrence of Sorex spp. and Microtus spp.

Signs of possible amelioration of Pleistocene conditions appeared in the stadial deposits for the first time in Bison Chamber with the occurrence of Notiosorex crawfordi, Perognathus sp., and Pitymys ochrogaster. Thomomys talpoides made its last appearance there. These changes may be an artifact, in that Loc. 6 and Loc. 4 have been well studied, while Loc. 22 and Loc. 54 have not. If real, the fauna implies that ground cover was becoming more scanty. Any such change must have been subtle, however, for other microtines and shrews remained.

Definite signs of warming summer temperatures appeared in the Human Corridor fauna, however, with the appearance of Sigmodon. Occurrence of Dipodomys cf. ordi there strongly suggests cover was diminishing. The slightly later sites of

Stalag 17 and TT II show <u>Neotoma</u> <u>floridana</u> and <u>Sigmodon</u>,
respectively. <u>N.</u> <u>floridana</u> indicates a summer-peak precipitation pattern was present, and presence of <u>Cryptotis</u> also
suggests an eastern-type climatic influence.

This eastern-type climatic regime appears to have been an
important factor in latest Pleistocene and, probably, the
early Holocene in the Southwest. It likely is represented in
Howell's Ridge Cave (<u>Cryptotis</u> <u>parva</u>) and the Isleta Caves
(<u>Sylvilagus</u> <u>floridanus</u>), and may have been the factor allowing <u>S.</u> <u>floridanus</u> to displace <u>S.</u> <u>nuttalli</u> from southern New
Mexico (with mid-Holocene restriction of <u>S.</u> <u>floridanus</u> to the
highlands of the region).

The Holocene fauna from the Entrance Chamber suggests in
common with evidence from other early Holocene sites and the
woodrat midden record that a considerably different climatic
regime was present during part of the time span than is typical of today. Specifically, the presence of <u>Neotoma</u> <u>mexicana</u>
--a taxon apparently absent from the rest of the Dry Cave
record--indicates climatic conditions different from those of
today, and a few <u>Microtus</u> seem to have hung on at this relatively low elevation after the end of the Pleistocene.

The Dry Cave sites allow some estimation of how likely it
is that rapid climatic fluctuations have swept several separate vegetative zones quickly through an area, mimicking
contemporaneity of those zones. The likelihood seems low.
Interstadial faunas seem relatively invariable and complex
for long periods of time--in excess of 12,000 years. The
Animal Fair progression makes clear that there has not been
wholesale mixing of faunas, but instead a transition from one
fauna not duplicated anywhere today to a second, also undup-
licated now. A period of rapid change at the end of the
Pleistocene is likely and a similarly rapid change at the

beginning of the full-glacial stade is possible. The relatively continuous nature of the later Dry Cave deposits and their dates make it apparent that the complexity of faunas either is real or must have been the result of massive changes over extremely short spans of time (measured in decades or scores of years). In my judgement, longer-term fluctuations would be revealed in the stratigraphic sections.

This evidence plus the generally consistent occurrence of the several vegetative zones throughout the suite of Interior Division sites makes it very unlikely that the observed complexity of Pleistocene faunas is an artifact.

VI. SUMMARY AND CONCLUSIONS

Excepting the presence of extinct forms, Sangamonian faunas
suggest conditions similar to those of today in most areas,
only revealing increased warmth over present conditions in
the most northern sites.

Wisconsinan-age sites preceding the late Wisconsinan full
stadial are too poorly dated to be correlated with oxygen-
isotope episodes of the marine record. Consequently, it is
unknown at this point which of the various climatic regimes
are represented by the faunas. Although referred to here as
interstadial faunas, in actuality, mid-Wisconsinan stadial or
transitional times could as well be represented. These sites
show conditions more like those of the present than those of
the late Wisconsinan full stadial, but nevertheless differed
from those of the present in several notable ways. Greater
effective moisture was available and at least the low temperature extremes of today's winters were absent. Several now-
extinct small vertebrates occurred into late mid-Wisconsinan
times, but apparently did not survive the late full glacial.
The unsure datings, the limited number of sites, and their
restricted geographic distribution result in almost over-
whelming ignorance regarding the basic features of times preceding the late Wisconsinan full glacial.

In contrast, full-stadial conditons in interior western
North America are revealed in broad outline, if not in detail. Plants and animals retreated at the northern and upper

elevational limits only moderately--in elevation, probably only some 600 to 700 m on the average. In contrast to the relatively small contractions seen at upper boundaries of biotic ranges, lower limits expanded drastically, probably averaging close to 1500 m in elevation. The result was a vast overlapping of ranges of organisms now separated by elevation or latitude, forming biotic complexes occurring nowhere today. The mechanisms driving this biotic complexification appear to have included more equable climatic conditions due to absence of the temperature extremes typical of the late Holocene, increased effective moisture due to cooler average temperatures and changes in patterns of precipitation, and increased slope-exposure effects on temperature and moisture.

The resulting biotic complexes can be conceptualized as 1) a northern and highland association (Northern Highland Zone) of tundra, boreal forest, sagebrush, and grassland elements, descending to an elevation of about 6050 m, equator equivalent (m ee); 2) a lower zone (Middle-elevation Savannah) of similar makeup, but lacking tundra elements and, in the lower half, with a woodland element, descending to about 5000 m ee; 3) a Sagebrush Steppe-woodland Zone descending to about 4550 m ee; 4) a Steppe-woodland Zone, lacking sagebrush-associated elements, with a lower boundary at about 4350 m ee or below; and 5) possibly a low-elevation Steppe Zone at the lowest elevations within the region.

The increased vegetational complexity, with the accompanying climatic features, allowed vertebrates now widely separated geographically to occur sympatrically or separated only microgeographically. All zones appear to have had relatively open areas present, though whether due to patchiness of tree

growth or to a savannah-type arangement in the upper zones is
uncertain.

Regarding the extinct megafaunal taxa, some important geographic and ecologic distributional data appear. Several
forms, present in the interior in mid-Wisconsinan times,
apparently became restricted to the West Coast Division and
to the extreme eastern portions of the study region during
full-glacial time. These forms included the edentates other
than †Nothrotheriops, the raccoon, the peccary †Platygonus,
the sabertooth †Smilodon, and possibly the gray fox. Mastodons were absent from the interior during the full glacial
except for a few relictual (?) populations in the extreme
southeast. †Capromeryx was a southern genus as probably was
†Euceratherium. Eastern species that today do not cross the
Northern Great Plains reached the region only along riparian
stringers on the eastern edge of our region. Major movements
within the region were southerly and southeasterly, with
Northern Plains and Great Basin forms reaching to the Mexican
border area in the southeast.

In the West Coast Division, differences from today were
less extreme than in the Interior Division, seemingly having
consisted mainly of movements downslope and to the south--the
degree of biotic mixing seen inland was not reached. The
lesser degree of change may reflect the already relatively
equable climate of the region, so that the changes that did
occur were primarily due to cooler air temperatures and some
exaggeration of slope effects.

The approach to the end of the Pleistocene was marked by
warming temperatures and shifts of precipitation patterns.
Evidence from southeastern New Mexico suggests that summer
precipitation became more important, with several eastern
species in that area moving westward into southern New Mexico

and some, probably, into southeastern Arizona or even central Arizona.

The end of the Pleistocene was marked by megafaunal extinctions and by climatic and biotic conditions more similar to those of the late Holocene than any recorded for the Late Pleistocene. Nevertheless, the extreme climatic conditions of the later Holocene and achievement of thoroughly modern biotic ranges occurred only sometime after about 8000 BP.

APPENDIX 1. AVIAN AND MAMMALIAN TAXA

Treated here are avian and mammalian taxa about which special comment seems desirable. Included are all the extinct mammals, those living taxa used in interpretation of the paleoecology, and extant taxa whose Pleistocene occurrences are of particular interest.

At the end of the appendix, several as yet undescribed species are briefly commented upon.

†Acinonyx trumani. American Cheetah.

This cat ranged from the bottom of our Middle-elevation Savannah Zone into the lower Northern Highland Zone, thus essentially being a boreal animal. Evidently a fast runner, it probably was an inhabitant of open country (Martin et al., 1977), presumably grassland or shrub-steppe.

Aegolius funereus. Boreal Owl.

This owl is a far-northern taxon normally breeding in Canada and Alaska and wintering no farther south than northern Montana in our region. It is recorded casually to as far south as Gunnison Co., CO (AOU, 1957). Ginn (1973:171) characterized this bird as one of "northern coniferous forests. Found especially among spruce, it occurs also in mixed forests among pine, birch and poplar."

Alces alces. Moose.

Moose are northern creatures closely associated with water in boreal forest situations.

Antilocapra americana. Pronghorn.

Only one species of the genus is recognized. Historically, pronghorn ranged from the Canadian prairie provinces south well into Mexico and from the West Coast to the eastern margin of the Great Plains. Morphologically it is a grazer, with well developed hypsodont cheekteeth. Dietary studies, however, indicate a majority of its food comes from forbs rather than grasses (O'Gara, 1978).

The Pleistocene record shows that it was basically as wide ranging during the Pleistocene as it is today, both geographically and elevationally. Misidentifications between it and its relatives, particularly Stockoceros, are to be expected, since only presence of the horncores allows easy identification between species of similar size.

Aplodontia rufa. Sewellel, or Mountain Beaver.

This large rodent is an inhabitant of the humid northwestern forests from northern California to British Columbia. Occurrence in the cave deposits of northern California indicates descent of their habitat to a lower elevation than today.

†Arctodus simus. Giant Short-faced Bear.

†Arctodus appears to have been basically an animal of the flats and foothills, although it is known from Potter Creek Cave in northern California. Kurten and Anderson (1980) commented on its probable unusual fleetness for a bear. Thus it may have been an animal of open country, utilizing rougher terrain primarily for denning. As pointed out by Kurten and

Anderson, occurrence with the grizzly bear (Ursus arctos) is rare in our area.

It is not clear that this genus was present at Conkling Cavern, and the Burnet Cave record is based only on the large size of several ursid vertebrae.

†Bassariscus sonoitensis. Sonoita Ringtail.

No new information beyond that recorded by Kurten and Anderson (1980) is added other than to point out that both sites from which the species is known (Papago Springs Cave, San Josecito Cave) are in the Sagebrush Steppe-woodland Zone elevationally. San Josecito Cave, in southern Nuevo Leon, Mexico, is so far from our area that it is difficult to suggest what validity zonal designation has, however.

†Bison antiquus. Ancient Bison.

A number of species of bison, all extinct except Bison bison, have been reported from our area and time span. J. N. McDonald (1981) has been the most recent reviewer and is followed here.

This species, ancestral to Bison bison, is considered by McDonald to include †B. occidentalis of other authors. It was widespread within the Wisconsinan, from as far south as Nicaragua north into Canada and from the West Coast to the East Coast (McDonald, 1981). McDonald suggested that records from the central and northern parts of our area date mostly from late Wisconsinan to middle Holocene. Our suite of sites indicates it occurred into the Northern Highlands Zone and to the bottom of the sagebrush zone. However, numerous lower-elevation sites, such as Ventana Cave and the Lehner Site in southeastern Arizona, recorded bison only to genus--many of these probably were †B. antiquus.

Bison may be instructive in regard to the grazing habit
and habitat. The living species reached its peak on the open
Great Plains grasslands, but the northern subspecies, B. b.
athabascae, "is known from the western boreal forest and bor-
dering ecotones of Canada north to Alaska." The "primary
range of the subspecies coincides with relatively extensive
parklands" (McDonald, 1981). Thus an open parkland or savannah habitat is not to be unexpected. Further, McDonald noted
seeing individual B. b. bison browse on woody plant foliage
and stems.

†Bison latifrons. Giant Bison.

This species apparently did not make it quite to the late
Wisconsinan full glacial, being last recorded within our
sites from the Rainbow Beach fauna (dates of 21,500 and
31,300).

Brachylagus idahoensis. Pygmy Rabbit.

This smallest of the living North American rabbits might
as aptly be called the sagebrush rabbit, for it occurs in the
cool sagebrush desert of the northern Great Basin and appar-
rently relies heavily on tall sagebrush for cover and food.
This also is an area of winter-dominant precipitation.
Pleistocene occurrence of pygmy rabbits south and east of its
present range is best interpreted as indicating well developed sagebrush stands with a cold-season precipitation peak.

†Camelops hesternus. Yesterday's Camel.

This species appears to have been the common large camel
of our time and region. Two other members of the genus have
been recorded (†C. huerfanensis and †C. minidokae), both at
American Falls, but Hopkins et al. (1969) considered the

identification of the camels from this site as very tentative.

The genus, and probably the species, was nearly ubiquitous, occurring from our highest site (Jaguar Cave) to
Schuiling Cave in the stadial faunas and seemingly was as
widespread during earlier times. It did not occur in our
lowest three sites in the Inland Division, but this well may
be due to sampling error. It apparently survived into Folsom
times (Walker, 1982b). On the West Coast, it was absent from
the higher sites, but in the Inland Division occurred from
the flats through the foothills and well into the more mountainous areas.

The hypsodont dentition suggests it was a grazer, but
possibly utilization of harsh, woody shrubs would call for
the same dental characteristics. Among living herbivores,
the domestic goat (Capra hirca) appears essentially as hypsodont as its close relative, the domestic sheep (Ovis
aries), but its proclivity for subsisting on a wide variety
of seemingly unpalatable foods is well noted in folklore.
Camels, and/or other Pleistocene large herbivores, may have
been adapted for feeding on various shrubs, such as sagebrush, which seems under-utilized by the current fauna.
Martin and Guilday (1967) pointed out that dromedary camels
introduced into the Southwest utilzed such shrubs as creosotebush and mesquite.

†Canis dirus. Dire Wolf.

This wolf-sized canid occurred, usually in moderate or
small numbers, throughout our time span and geographic area.
Most occurrences among our sites were in the foothills and
low mountains, though it occurred both onto the flats and into more mountainous areas on the West Coast. Likely denning

sites account for much of the distribution of the remains.
Among extinct species, only †Panthera leo atrox ranged as
widely elevationally (ee), though possibly the ability to
identify accurately to species some of the wider-ranging herbivores would show equal dispersion.

Canis lupus. Gray Wolf.
    Although well represented throughout the late-glacial
sites, C. lupus was recorded in Sangamonian sites only in
Canada (Fort Qu'Appelle and Medicine Hat). Remains also were
rare in the interstadial sites, but the distribution pattern
(Fossil Lake and several Dry Cave sites) suggests the rarity
is an artifact of sampling error.

†Capromeryx minor. Diminutive Pronghorn.
    Although many of the very small pronghorns are identified
only to genus, most or all probably belonged to this species.
This seems to have been predominantly a southern, relatively
low elevation form during the late glacial. A queried occurrence at Smith Creek Cave was some 1150 m ee above the next
highest record. Disregarding this record as probably erron-
eous, occurrences were limited to below 5000 m ee in southern
California, Arizona, and New Mexico into Mexico. An uncertain identification is available from the San Francisco Bay
area. Interstadial and Sangamonian records likewise seem
similarly limited.
    The pluvial occurrences would place this creature in the
Sagebrush Steppe-woodlands and the Steppe-woodlands (possibly
below at Jimenez Cave). Remains were common in the foothills
and low mountains, but absent in the high mountains and rare
in flatland situations.

Castor canadensis. Beaver.

Beavers occur nearly everywhere that food and sufficient water for protection are available, and thus tell little beyond the presence of such conditions. Although usually in forested areas, I have observed beaver sign along the San Juan River in New Mexico where only shrubby willows and the like occur, so even a well developed riparian forest cannot be assumed.

Centrocercus urophasianus. Sage Grouse.

As implied by the common name, the sage grouse is associated closely with sagebrushes of the genus Artemisia, and particularly with big sagebrush (A. tridentata). Historically, these grouse have occurred only within the region where sagebrushes form extensive stands (Johnsgard, 1973). Thus the Great Basin and northeast into the Northern Plains and, sporatically, south into northern New Mexico is its historic range.

Migrational movements generally are less than 100 miles and tend to be movements between higher and lower elevations (Johnsgard, 1973).

In view of the present association of sage grouse and sagebrush and that the major distribution of stand-forming sagebrushes today are in areas tending to receive appreciable amounts of winter precipitation, past occurrences can be viewed as evidence for both such vegetational and climatic characteristics.

†Cervalces scotti. Stag-moose.

Most occurrences of this large moose-like mammal occurred in the East, but it was recorded from Qu'Appelle within our area. It probably was an inhabitant of aquatic environments.

Cervus elaphus. Wapiti (Elk).

Historically, wapiti occurred not only in forested areas, but also to some degree out onto the plains. In the south, occurrences were mostly in the mountains. Guthrie (1968) pointed out that considerable grass occurs in the diet.

Most of the stadial records of _Cervus_ were in the Northern Highlands Zone in the northern part of our area, but it was recorded as low as Papago Springs Cave (_Cervus_ sp.) and Williams Cave (_C_. cf. _elaphus_) in the south, both in our Sagebrush Steppe-woodland Zone (Williams Cave included Holocene material, however). The rarity of occurrence in the south suggests the animal was not common and that the possibly larger historic numbers reflect a niche expansion made possible by extinction of competitors.

Clethrionomys gapperi. Boreal Red-backed Vole.

This vole tends to occur in relatively closed boreal forest habitats, seemingly less requiring of the moderate to heavy grass growth generally inhabited by voles of the genus _Microtus_. The fossil occurrences were in the Northern Highlands and Middle-elevation Savannah zones (Little Box Elder and Wilson Butte caves, Agate Basin Site).

Conepatus mesoleucus. Hog-nosed Skunk.

Hog-nosed skunks are dominantly southern animals today, although their range does extend north into Colorado. Ecologically, they are most common in Lower Sonoran areas, but do extend upward frequently into woodlands and, in desert ranges, into marginal Transitional Life Zone situations.

Wisconsinan occurrences were in interstadial and stadial deposits in the Guadalupe Mountains area and at San Josecito Cave in Mexico.

Cryptotis parva. Least Shrew.

Unlike the species just considered, the least shrews have the bulk of their range south of the boreal forests and east of our area. Although deciduous forests may be inhabited, more typically grassy clearings or low vegetation types with abundant grass is the norm. Porter (1978) showed that the present western boundary of the range likely is governed by moisture requirements. Its presence in Pleistocene faunas within our area likely implies rather open habitat with plentiful grasses was present.

Fossil occurrences were at Little Box Elder Cave (Cryptotis sp.), where it may represent a pre-Altithermal relict, and from southern New Mexico and Chihuahua. The pattern in the south suggests the historic pattern of such grassland forms as Cynomys ludovicianus, but of a more mesic aspect. The southern occurrences were within the Middle-elevation Savannah and the Sagebrush Steppe-woodlands.

Cynomys. Prairie Dogs.

The generally parapatric ranges of living U.S. prairie dogs strongly suggest competition as important in determining which species occupies a given area within their common range. The white-tailed group, Cynomys (Leucocrossuromys), centers west of the black-tailed prairie dog (Cynomys ludovicianus). The latter occupies the grasslands of the Great Plains and the somewhat more arid projection of these grasslands across southern New Mexico into southeastern Arizona. The white-tailed group inhabits the higher-elevation open grasslands to the west and extends into open woodlands and forest clearings; Armstrong (1972) reported them to 3650 m in Colorado. These western species, represented in the Pleistocene record by the northern C. leucurus and the southern C.

gunnisoni, may be less demanding of large, unobstructed areas than is C. ludovicianus. The latter occurs in areas of primarily summer precipitation or of approximately equal warm and cold season precipitation, whereas the other species tend to go from the latter conditions to areas with winter-dominant precipitation.

Fossil occurrences of the white-tailed group fell mostly into their approximate present range, but also were noted in stadial faunas in southeastern New Mexico beyond their current range. The interstadial sites recorded C. ludovicianus in this area. The specimen tentatively identified by Harris (1970b) as C. cf. gunnisoni from ca. 5 miles S Alpine, Brewster Co., Texas, is reassigned to C. cf. ludovicianus. C. ludovicianus also occurred north of its present range in presumed Sangamonian times and west (to Ventana Cave) in the full-glacial and post-Wisconsinan.

†Dasypus bellus.   Beautiful Armadillo.

†Dasypus bellus is reported only from Blackwater Draw and appears to have been an eastern form following river-valley growth up the Brazos drainage. Slaughter (1961:313) hypothesized that presence indicated relatively mild winters ("no more severe than is found in North Central Texas today") and with at least a moderate amount of precipitation ("rainfall was probably more than twenty inches per annum").

†Desmodus stocki.   Stock's Vampire Bat.

Today, vampire bats (Desmodus rotundus) find their northern limits well south of the U.S. border. Kurten and Anderson (1980) noted that present occurrences are not north of the 10°C winter isotherm. Since their prey is common to the north today, the implication is that temperature is limiting;

more moisture probably is required than occurs in the northern hot deserts, too.

The Pleistocene form likely was merely a D. rotundus
characterized by larger size and concomitant changes, and it
seems likely that it was governed by much the same limitations as the modern form.

†Desmodus stocki was recorded from two full-glacial sites,
Potter Creek Cave in California and San Josecito Cave in
Nuevo Leon. It, or possibly the living species (D. rotun-
dus), also was recorded from the southern Big Bend of Texas
(Ray and Wilson, 1979). The age is unknown, but presumed to
be Wisconsinan.

Dicrostonyx torquatus. Collared Lemming.

This is an animal of arctic tundra, not reaching as far
south as the latitude of southern Hudson's Bay at present.
Pleistocene occurrences in our area presumably indicated
tundra-like vegetation, and it is used as a tundra indicator.

Didelphis virginianus. Virginia Opossum.

This species is an eastern and southern form that in the
Late Pleistocene appears to have been one of several taxa
moving up riparian corridors through the plains to the eastern edge of our area (Blackwater Draw).

Dipodomys. Kangaroo Rats.

Members of this genus are widespread today, mostly in
desert and grasslands, but a few in chaparral and woodland
habitats. One species, D. ordi, extends well into the
Canadian Prairie Provinces.

At least four species were recorded from our time span,
but remains were surprisingly scanty considering the present

distribution. The West Coast identifications, when given to
species, are of forms occurring nearby today.

Dipodomys ordi. Ord's Kangaroo Rat.

Ord's kangaroo rat occurs throughout the more arid region
of our area except for a wide strip along the West Coast and
from Washington to central Montana. The feature of its ecology of value to paleoecology lies in its requirement of an
open microhabitat--heavy grassy growth, brush, and closed
woodland is avoided, although pinyon-juniper woodland on a
sandy substratum and with a fairly well developed understory
of grass and herbaceous growth may be inhabited (Harris,
1963a). Presence of D. ordi in fossil faunas, then, implies
at least local areas with relatively sparse vegetation were
present. Absence in large faunas, where sampling error is
unlikely to be a factor, suggests there was a relatively
heavy vegetative ground cover.

Most supposed stadial occurrences of D. ordi were in sites
known to include Holocene material or suspected of including
pre-stadial fossils (Smith Creek Cave). The only reasonably
sure stadial occurrence was D. cf. ordi from the Human Corridor site in Dry Cave. The absence of identified ordi-sized
Dipodomys from earlier stadial sites at Dry Cave is considered significant--the taxon probably was not in the area.

Dipodomys spectabilis. Bannertail Kangaroo Rat.

The large D. spectabilis seemingly remained in at least
part of its present range during the full glacial, having
been present in central and southeastern New Mexico. It also
was recorded from the Dry Cave interstadial deposits, but the
specific identification is not entirely satisfactory. It
presently is a Lower and Upper Sonoran desert and grassland

form extending to the lower border of the pinyon-juniper woodland. Its range is limited to Arizona, New Mexico, and western Texas in the U.S., continuing on south into Mexico.

Eptesicus fuscus. Big Brown Bat.

This is one of the more commonly recorded fossil bats, probably in part due to its ease of recognition. All occurrences probably were stadial. The fossils average larger than populations in the same regions today, presumably as an adjustment to generally cooler temperatures. The taxon formerly was known as an extinct species, †E. grandis, but rather obviously should be recognized as E. fuscus (Guilday, 1967).

†Equus. Horses.

Despite the long history of study of American Pleistocene horses, there remains a wide variety of views on relationships and the number of recognizable species. Most sites in our area recorded from one to three species, commonly differing in size. Unfortunately, differences in nomenclature and in the concept of the characters of a given species make generalizations dangerous, and changes in nomenclature and concepts through time add a third dimension of possible error.

In my own understanding of the situation in the Southwest (Harris and Porter, 1980), two species (†E. conversidens and †E. niobrarensis) were widespread in the Inland Division and a third (†E. occidentalis) became increasingly more common from east to west, having been the common horse on the West Coast. Two other forms were more limited: †E. tau, primarily southern, and †E. pacificus, northwestern. An undescribed zebra, the †Species A of Harris and Porter (1980), is known from a single site of uncertain stadial age.

As understood by me, †E. occidentalis, †E. conversidens, and †E. tau (and also †Species A) had in common a lack of infundibulae in the lower incisors. Thus they would be considered by some to be members of the subgenus †Amerhippus. The three graded in size from large to small, respectively. The specimen from Dry Cave recorded by Harris and Porter (1980) as †E. scotti probably represented †E. occidentalis, making these three taxa contemporaneous during the Dry Cave full glacial. Several other sites appear to have had associations of †E. conversidens and †E. niobrarensis, including Burnet Cave, Isleta Cave No. 2, and Gypsum Cave (where †E. conversidens was consistently smaller than in the more southeastern sites).

Other nominal taxa recorded in the literature include E. caballus, †E. excelsus, †E. laurentius, †E. fraternus, †E. giganteus, †E. (Asinus), †E. (Hemionus), and †E. (Amerhippus). Some of these may indeed represent species other than those mentioned above, but the uncertainty is great.

†Equus conversidens.   Mexican Ass.

†Equus conversidens is recorded widely in the literature from the northern limits of our area in Canada south into Mexico (San Josecito Cave). It is difficult to judge if all such records pertain to the same taxon, but probably many do. The very small hooves and relatively long metapodials suggest it inhabited open areas with a firm substratum, though the hoof size also suggests the animal was at home in rugged terrain.

†Equus niobrarensis.   Niobrara Horse.

†Equus niobrarensis was a moderate-sized horse with broad hooves. This probably is the correct name for some of the

Wisconsinan horses recorded in the literature as †E. scotti
(Harris and Porter, 1980), but whether the two are indeed the
same taxon is unclear at present. Some occurrences listed in
the literature as E. caballus probably pertain to this species. Occurrence was throughout the Interior Division from
Canada to Trans-Pecos Texas (Salt Creek) and southern New
Mexico. The relatively broad hooves may have been adapta-
tions for a softer substratum than that utilized by †E. conversidens, and this horse may have been most common in riparian situations or the more heavily wooded sections. The distribution is consistent with a soft substratum in that the
lowest occurrence ee (Dark Canyon) coincided with the hypothesized major habitat change from Sagebrush Steppe-woodland
to Steppe-woodland. This should not be taken too seriously,
however, considering the number of Pleistocene specimens
identified to the generic level only.

†Equus occidentalis. Western Horse.

   †Equus occidentalis was a common horse on the West Coast,
but relatively seldomly reported from inland. Harris and
Porter (1980) tentatively reported †E. occidentalis from the
Dry Cave interstadial, but were unsure of the identification
because of dental characteristics not mentioned in the literature. Since then, I have examined specimens from McKittrick; these agree in the presence of hypostylids on DP_{3-4}
and other dental features, and there seems no difficulty in
assigning the Dry Cave interstadial specimen to this species.
Occurrence in the stadial deposits is less sure, being dependent on the similarity of a partial skeleton to that of †E.
occidentalis (Human Corridor) and a lower incisor without
infundibulum that appears to be too large to represent †E.

conversidens. Identifications in the older literature, such
as for Sandia Cave, need confirmation.

†Equus pacificus. Pacific Horse.

†Equus pacificus was another large horse, near the size of
†E. occidentalis, but infundibulae were present in the lower
incisors. Records are from the West Coast, from Fossil Lake,
and from Lake Lahontan. Confusion between the two large species is possible for the latter site. †E. pacificus also was
recorded from a single site (Ingleside) from east of our area
(Lundelius, 1972a).

†Equus tau. Small Stilt-legged Horse.

†Equus tau was a small, pronouncedly stilt-legged horse
recorded in our area from Papago Springs Cave and, less certainly, from Arizpe and Burnet Cave. The Burnet Cave record
is based on teeth that appear to show the effects of semi-
digestion, decreasing certainty of identification and also
suggesting that the remains were transported in from some
distance.

†Euceratherium collinum. Shrub-ox.

†Preptoceras is treated here as a synonym of this extinct
bovid. The elevational distribution of this form was limited, ranging from about 4800 m ee (San Josecito Cave, Conkling
Cavern) to 5682 m ee (Abiquiu). All occurrences were in
foothills or relatively low mountainous terrains. Geographically, it ranged from California and Nevada (Mineral Hill)
southeasterly to northern New Mexico and Trans-Pecos Texas
(Hord Shelter) and south into Mexico (San Josecito Cave).
Sites were within the Middle-elevation Savannah and the Sagebrush Steppe-woodlands.

Geococcyx californianus. Roadrunner.

Harris and Crews (1983) recently have hypothesized that the nominal species †G. conklingi, named mostly on the basis of large size, is a subspecies of the California roadrunner and that size differences between Pleistocene populations reflected different temperature regimes. As hypothesized, large body size characterized populations adapted to cold winters and cool summers (†G. c. conklingi), while body size similar to that of modern populations characterized birds adapted to summers with high temperature extremes (C. c. californianus).

Geomys. Eastern Pocket Gophers.

These fossorial mammals are predominantly grassland forms. Fossil occurrences were within the present range of the genus except for the Hand Hills site, in Alberta. This site, recorded by Harington (1978) as probably Irvingtonian or later, is considerably to the north and west of the current range (nearest modern records are in extreme south-central Manitoba). Cynomys ludovicianus also was recorded from the site and possibly the site was Sangamonian.

†Glossotherium harlani. Harlan's Ground Sloth.

Kurten and Anderson (1980) characterized this large sloth as a grassland species. Its record in our area seems consistent with this habitat. Full-glacial occurrences among our sites were limited to the eastern plains and the West Coast. However, it appears to have been much more wide-spread during Sangamonian and interstadial times.

Gulo gulo. Wolverine.

Wolverines generally occur in tundra or heavy boreal

forest habitats. They have been used here as a marker for
presence of northern-type forest. They are known to range
eastward into the plains in Nebraska, however (Jones, 1964).

†Hemiauchenia macrocephala. Large-headed Llama.
    This large, slenderly-built llama was characterized by
Kurten and Anderson (1980) as a plains-dwelling grazer.
Inspection of site distributions containing †Hemiauchenia,
however, shows approximately equal division between flatland
sites and sites in the foothills and low mountains. The dentition was less hypsodont than in most living advanced grazers, and it seems likely to me that grasses were not the
primary food.
    Although identifications to the species level from stadial
sites are limited in elevational range, the genus was recorded from all of our zones except Steppe in the Inland Division. The genus seems to have been somewhat uncommon on the
West Coast. Pre-stadial occurrences were mostly northern or
on the West Coast, but this may reflect sampling error.

†Homotherium sp. Scimitar Cat.
    This was a sabertooth cat with relatively short upper
canines. It was reported from the American Falls Sangamonian
fauna, but most occurrences within our time span were to the
east of our area.

Lagurus curtatus. Sagebrush Vole.
    Sagebrush voles are strongly associated with big sagebrush
(Artemisia tridentata) and similar species, from the southern
portions of the Great Basin northeastward to the Northern
Plains of Montana and North Dakota, and north into Canada.
In many areas, the sagebrush habitat is more arid than habi-

tats utilized by most other kinds of voles.

Since the sagebrush seems to be of considerable importance to the vole for food, rather than the vole merely responding similarly as the sagebrush to climatic factors, Pleistocene occurrences outside the present range probably should be taken as strong evidence of the presence of sagebrush stands nearby.

Fossil occurrences at stadial sites in New Mexico (Atlatl Cave, Isleta Caves, Howell's Ridge Cave, at least five of the Dry Cave sites, and Dark Canyon Cave) document its presence far to the south and east of its present range. Dark Canyon Cave is the lowest site ee at which it has been recorded. It also occurred at high-elevation sites.

Lepus. Jack Rabbits.

Jack rabbits were nearly ubiquitous throughout our area. Identification is extremely difficult even with good material for most species.

Lepus alleni. Antelope Jack Rabbit.

Lepus alleni is a Sonoran Desert form. Schultz and Howard (1935) referred a partial dentary from Burnet Cave to this species. I have examined the specimen and conclude it is not separable from other large species of Lepus (L. californicus and L. townsendi).

Lepus americanus. Snowshoe Hare.

The snowshoe hare is a small boreal form extending southward in the major mountain ranges of the West. It generally is associated with heavy spruce-fir forest up into the alpine zone. Within the forest, "optimal cover . . . is stands of brush" (Armstrong, 1972:88). It is considered to be a marker species for spruce-fir forest.

It was rare as a fossil, occurring at Jaguar, Potter Creek, and Samwel caves.

Lepus californicus. Black-tailed Jack Rabbit.

Lepus californicus is widespread at lower elevations today. It inhabits desert and grasslands primarily, but extends into open woodland and even marginally into ponderosa pine forest on occasion. It is more southern in its distribution than the white-tailed jack rabbit (L. townsendi), extending north to southern South Dakota in the plains and, west of the Rockies, to extreme southwestern Montana and southern Washington. L. townsendi extends from central Saskatchewan south to extreme northern New Mexico.

These two species are difficult to identify, although L. townsendi averages slightly larger and shows other average differences. Basioccipital breadth versus bullar breadth seems to separate them well, but this measurement requires an intact occipital region. An imperfect separation is possible by tibiofibula measurements. The distal breadth of seven L. townsendi ranged from 13.7 to 15.9 mm, all but one >14.3 mm. The same measurement on 13 L. californicus ranged from 12.6 to 14.1, with 10 of the 13 being <13.7 mm. A similar situation is found with proximal breadth: L. townsendi, 17.5 to 19.9 mm, with six of the seven >17.9 mm; L. californicus, 15.9 to 17.6 mm with all but one being <17.5 mm. Overlap would be expected to increase with a decent-sized sample, but where a sample of fossil specimens is available, the bulk often falls clearly into one or the other size range. Other subsidiary data may aid. For example, L. californicus tends (but with exceptions) to have heavily crinkled posterior walls to the lower cheektooth reentrant angles and L. town-sendi tends to have little crinkling.

Lepus californicus may have been entirely absent from the stadial deposits of the Interior Divison; identifications in the literature generally have been based on inappropriate materials or were from sites known to have included Holocene materials.

The interstadial deposits at Dry Cave contained this species.

Lepus townsendi. White-tailed Jack Rabbit.

This large hare "is usually characteristic of broad, open plains, but it follows open country up mountain slopes to altitudes varying from 10,000 to 12,000 feet . . . ." (Nelson, 1909:72-73). The distribution is the interior two-thirds of our area; it extends eastward across the plains and southward in the plains nearly to southern Kansas. Although overlapping to some degree with the black-tailed jack rabbit on the south, it seems to do best in cooler climate and some-what more mesic conditions than the latter. This jack rabbit appears to have been the common (and possibly the only) large hare in the Interior Division during stadial times.

Lynx canadensis. Canadian Lynx.

This northern relative of the bobcat is associated with boreal forest and is used as a marker species for that habi-tat. As a fossil, it occurred in the full-glacial deposits of the northern USA.

Macrotis californicus. California Leaf-nosed Bat.

The phyllostomatid bats are primarily tropical and semi-tropical forms currently entering the U.S. only along the

southern border. Macrotis occurs now in the Sonoran Desert
in the U.S., becoming fairly widespread farther south in Mexico. Its occurrence in the Pleistocene presumably indicated
mild winter tempertures. It was recorded from the undated
Cinnabar Mine locality in the Big Bend of Texas.

†Mammut americanum. American Mastodon.

Mastodons were relatively rare in the pluvial Interior
Division. Most of the presumed latest Pleistocene records
clustered in the Sandia Mountains of New Mexico (Sandia Cave,
Placitas, Tree Springs). It also was recorded from southeastern Arizona near Naco and at Lehner Ranch, but at Lehner
Ranch was dated at more than 30,000 BP (Mead et al., 1979).
West Coast records ranged from Sangamonian to late glacial
and the animal appears to have been moderately common and to
have roamed all types of terrain.

†Mammuthus. Mammoths.

As with horses, general agreement on nomenclature and
classification of mammoths is not yet here. As a result, it
is difficult to reconcile identifications made by different
workers. The wooly mammoth (†M. primigeneus) appears recognized by most workers, and most localities were far to the
north in our area. Presumably its identification at Potter
Creek Cave referred to a different species, however.

More southern forms pose the major problems. Kurten and
Anderson (1980) recognized the late Rancholabrean mammoths as
†M. jeffersoni and limited †M. columbi to its middle Irvingtonian to middle Rancholabrean predecessor. Most workers in
the West, however, consider the mammoths of latest Wisconsinan time as †M. columbi. Some workers in our area recognize a
separate species, †M. imperator (considered a synonym of †M.

columbi by Kurten and Anderson). The problem is further com-
plicated by the fact that most mammoth material is unassign-
able regardless of the concept held by a particular worker.

Within our basic suite of sites from the Interior Division, †Mammuthus sp. was limited to below about 6200 m ee and
to above about 4550 m ee. At its lower margin then, it par-
ticipated in the dropout at the base of the Sagebrush Steppe-
woodland. On the West Coast, it ranged throughout our elevational span.

Marmota flaviventris. Yellow-bellied Marmot.

This species is one of the most widely recovered extant
species from Wisconsinan deposits. Since many occurrences
were outside of its current range, it frequently has been
used as a basis (often the sole basis) of ecological reconstructions. In part, however, the reconstructions have been
based on misreadings of the current ecology of the animal
(Harris, 1970a). In the southern portions of its range
(south to northern New Mexico), it generally is a high alti-
tude creature. Its natural occurrence at much lower elevations in parts of its range and its viability at relatively
low elevations even in the south in situations where irriga-
tion runoff provides ample green fodder (Harris, 1963a; Durrant and Dean, 1961), however, indicate that temperature
requirements are not the critical factor. As discussed by
Harris (1970a), availability of green food plants during the
active periods of this hibernator seems to be the key to its
distribution. Even occasional years where such food was unavailable would eradicate the animal.

Aside from food requirements, marmots normally are associated with rocky areas where burrows may be dug between
boulders to protect from excavation by predators; tree roots

may be utilized similarly and the animal is known from fossil
occurrences where many miles of non-rocky terrain must have
been traversed during colonization (e.g., Baldy Peak Cave in
the Florida Mountains of southwestern New Mexico). Absence
today from seemingly suitable mountain ranges beyond its present range despite its occurrence in those highlands during
the Pleistocene implies unsuitable conditions some time in
the past (during the Altithermal?) with inability to recolon-
ize across lowland aridity barriers.

Martes americana. Pine Martin.
   The pine martin is used as an indicator of spruce-fir
forest.

†Martes nobilis. Noble Martin.
   This extinct mustelid presumably inhabited spruce-fir
forest. Its Pleistocene distribution was south to northern
California, Utah, and northern Colorado.

†Megalonyx jeffersoni. Jefferson's Ground Sloth.
   Probably all the †Megalonyx specimens are assignable to
this species. Stadial specimens were from the West Coast and
Tule Springs, but earlier specimens occurred also in the
northern portion of our area.

Microdipodops megacephalus. Dark Kangaroo Mouse.
   This animal occurs today almost to the elevation of the
sole Wisconsinan occurrence at Smith Creek Cave. It seems
unlikely that it was a member of the stadial fauna, but probably a pre-stadial element, as hypothesized for Panther onca
from the site, or a Holocene intrusive.

Microtus.   Meadow Voles.

Voles of the genus Microtus basically are grazers, closely
associated with grasses or grass-like plants such as sedges.
Normally such growth must be heavy enough to provide conceal-
ment during feeding activities, and most species construct
runways.   Where two or more species are sympatric, competition tends to result in habitat segregation, at least during
times of population lows.   Under some circumstances, considerable overlap may occur.   Most arid and semi-arid regions
support no species except occasionally as relict populations
in especially favored sites.   Identification to species often
is difficult.

The mountain-dwelling meadow voles considered here tend to
form a series in terms of tolerance to aridity.   M. pennsylvanicus is most limited to hydrosere conditions; increasing
tolerance is shown in the sequence: M. montanus, M. longicaudus, and M. mexicanus.   Occurrence of up to three species
at a single fossil site (e.g., Howell's Ridge Cave) implies
there must have been considerable diversification of habitat
in the vicinity.

Microtus californicus.   California Vole.

This mouse is widespread within the West Coast Division
from Oregon into Baja California and in a wide elevational
range.   Some populations occur east of the Sierra Nevada,
including a relictual population in California near the
southern Nevada border (near Shoshone, 475 m).

As a fossil, it occurred within its present range in both
stadial and Sangamonian sites.   It also was identified tentatively from Tule Springs E-1, slightly east of its present
range.

Microtus longicaudus. Long-tailed Vole.

The long-tailed vole is extremely widespread in the mountainous West, extending farther in virtually all directions
than does M. montanus. Although it will inhabit hydrosere
communities if given the opportunity, it also does well in
drier meadows, forest-edge areas, and into forest itself.

As a fossil, it appears to have been limited in the south
to the southeastern New Mexican sites.

Microtus mexicanus. Mexican Vole.

This is the most arid-adapted of the mountain-dwelling
meadow voles. Typically it inhabits meadows in open ponderosa pine forest and mixed coniferous forest. Where one or
more other species of vole occurs with it, it tends to be
restricted to the drier sites. Pinyon-juniper woodland may
be inhabited (possibly mostly during population highs) if
suitable understory plants are available for food and cover.
Some evidence from archaeological faunas suggests the current
sparsity of such growth in many parts of the Southwest is a
result of historic overgrazing (Harris, 1977a); if so, Mexican voles may have been wider spread in such habitat in the
recent past. Its range barely reaches north to Utah and
Colorado, with most of its distribution in Arizona, New Mexico, and Mexico.

Fossil occurrences were concentrated within the present
general area now inhabited, but at considerably lower elevations and, if Slaughter's (1975) identification (cf.-level)
at Blackwater Draw was correct, far from the highland areas
presently occupied.

Microtus montanus. Montane Vole.

The montane vole, true to its name, is a predominantly

montane form and most common in hydrosere communities, though not limited to such. Under ideal conditions, it may inhabit areas as low as the upper woodlands. Except for a few mountain ranges in northern and western New Mexico and the White Mountains of Arizona, it is absent from these states, most of its range being to the north and west.

It was recorded from Sangamonian, interstadial, and stadial sites in the northern and central western USA, and from far to the south of its present range in stadial sites of New Mexico. The latter included sites from the Rio Grande Valley area (Anthony Cave, Shelter Cave) and to the west (Howell's Ridge Cave), but it has not yet been identified in the southeastern New Mexican sites. The complementary nature of fossil occurrences of M. montanus and M. longicaudus in southern New Mexico suggests competitive exclusion.

Microtus pennsylvanicus. Meadow Vole.

Meadow voles are widespread in boreal America and, to some degree, to the south. In the West, populations occur in mesic habitats in the Northern Plains and in both the Northern and Southern Rockies. Relictual populations are scattered south through western New Mexico to northwestern Chihuahua. Where such populations are at relatively low elevations, such as along the San Juan River in northwestern New Mexico or in Chihuahua, the occurrence is limited to sedge beds maintained by permanent water sources. Occurrence in hydrosere communities is the general rule also within the main range to the north (e.g., Armstrong, 1972).

Fossil occurrences were in Wyoming, New Mexico, and Chihuahua. Many of the southern sites are far removed from the nearest present day populations both geographically and elevationally. It appears to have hung on in the Rio Grande

Valley of New Mexico into Holocene times (Smartt, 1977).

Mustela erminea.   Ermine.
This small weasel generally inhabits boreal forest habitat, and is used here as a marker for such.

†Mustela reliquus.   Relictual Weasel.
This weasel was described by Hall (1960) from San Josecito Cave and has been recorded only from that site. Kurten and Anderson (1980) included it as a synonym of the long-tailed weasel (M. frenata) without comment.

Myotis.   Mouse-eared Bats.
A wide variety of these bats occur in the Southwest, decreasing in number of kinds to the north. Although of biological interest, their flying ability, tendency to migrate at least locally to and from hibernacula, and the difficulty in identifying incomplete specimens leaves them of relatively little value in paleoecologic research at present.

Myotis lucifugus.   Little Brown Myotis.
Late-glacial specimens tentatively identified from Dry Cave appear more typical in size and dentition to the more northern M. l. carissima rather than to the present nearby populations (M. l. occultus).

†Myotis rectidentis.   Straight-tooth Myotis.
Described from Lauback Cave in Texas, this species may have occurred also in two stadial sites in Dry Cave.

Myotis thysanodes.   Fringe-tailed Myotis.
The record from Isleta Cave No. 1 may have been Holocene.

A colony is known to have inhabited the cave within the last few decades and some of the specimens definitely are modern.

Myotis velifer.   Cave Myotis.

   This taxon includes †M. magnamolaris, described as extinct by Choate and Hall (1967). Dorsey (1977) concluded that M. velifer and †M. magnamolaris were conspecific in a paper apparently missed by Kurten and Anderson (1980). Pleistocene specimens seem to have averaged larger than those of the Holocene.

   Pleistocene records were within the present geographic range of M. velifer.

†Navahoceros fricki.   Mountain Deer.

   Within our area, this extinct deer occurred in the foothills and low mountains from Little Box Elder Cave in the north to San Josecito Cave in the south (there are other records in Mexico beyond our region, however). Between these extremes, the only records were from southern New Mexico: Burnet Cave, Cueva Las Cruces (?), Hermit's Cave, and Slaughter Cave. The Little Box Elder Cave record (Kurten and Anderson, 1980) seems anomalous in that other records were tightly clustered within the Sagebrush Steppe-woodland Zone, while Little Box Elder Cave was well into the Northern Highland Zone. The Little Box Elder specimen was not recorded in the original publication (Anderson, 1968).

Neotoma.   Woodrats.

   Some eight species of Neotoma have generally been recognized as occurring in our primary area; many are of potential importance for elucidating past ecology. Paleontological identification to species is difficult, although division to

subgroupings often is easier. Discriminating criteria, particularly of M_1, have recently been studied (Harris, in
press,a), and many of the identifications here are based on
that study. Commonly two or three species, and sometimes
four, may occur in close proximity, both in modern and fossil
situations, complicating interpretation of environments unless satisfactory samples are available.

Neotoma floridana. Eastern Woodrat.

    Presence of N. floridana seems well established in the
Stalag 17 fauna of Dry Cave on the basis of the anterior two-
thirds of a skull and two dentaries. There were probable
occurrences at Jimenez Cave and TT II (Dry Cave). The eastern woodrat now occurs east and north of these areas. The
dated Pleistocene occurrences were from near the end of the
Wisconsinan and may reflect an eastern precipitation pattern
and warming temperatures appearing near the end of the
Pleistocene.

Neotoma goldmani. Goldman's Woodrat.

    This small woodrat currently occurs in north-central Mexi-
co to nearly as far north as the Big Bend of Texas. Specimens from the Animal Fair site of Dry Cave have been identified as N. ? goldmani (Harris, in press,a). This may be the
taxon tentatively identified by Schultz and Howard (1935)
from Burnet Cave as N. lepida.

Neotoma lepida. Desert Wood Rat.

    This is a western species of desert and woodland. Most
published Wisconsinan records were from within its current
range. The major exception was at Burnet Cave, which, con-
trary to Kurten and Anderson (1980), is far to the east of

its present geographic range. I have elsewhere (Harris, 1970b) suggested the Burnet Cave specimen probably represented N. stephensi. However, the recent identification of N. ? goldmani from Dry Cave and the absence of N. stephensi from the fossil record of southeastern New Mexico suggests this may be the taxon represented at Burnet Cave.

The occurrence at Jimenez Cave (? stadial) was far outside the modern range, and the population differs in several ways from modern N. lepida (Harris, in press,a).

Occurrence at Smith Creek in late-glacial deposits is suspect. Although expectable in the area at present, the deposits were laid down in a Northern Highlands Zone environment far above other inland occurrences. A Holocene origin is probable.

Neotoma stephensi. Stephen's Woodrat.

This small Neotoma is an inhabitant of the Colorado Plateau woodlands. It appears more associated with junipers than is any other species (Harris, 1963a). In areas also occupied by other species of Neotoma, its midden often can be distinguished by the higher proportion of juniper clippings. Published records at Rampart and Vulture caves (cf.) may be of N. lepida.

Neotoma mexicana. Mexican Wood Rat.

Despite having the major portion of its distribution in Mexico, N. mexicana is predominantly a highland form in our area. It occurs from pinyon-juniper woodland up well into coniferous forest habitats. Generally N. mexicana extends to higher elevations than the white-throated woodrat (N. albigula), but does overlap with that species. In mountainous areas of proper elevation, N. mexicana often inhabits north-

facing slopes while N. albigula lives on adjacent southerly
slopes.

Tentative identifications at Isleta Caves and Dry Cave
(Harris and Findley, 1964; Harris, 1970b) are incorrect, N.
cinerea being the species represented. The species does not
seem to be associated surely with any full-glacial fauna.
Surprisingly, however, there were several occurrences in early Holocene deposits (the Dry Cave Entrance Chamber and the
Khulo Site of south-central New Mexico), suggesting the animal replaced N. cinerea in lowland situations in the Southwest at the end of the Pleistocene and hung on for some time.

Neotoma cinerea. Bushy-tailed Woodrat.

This rodent strongly prefers areas of "vertical cracks,
crevices, and 'chimneys' as den sites. Natural caves . . .
also are utilized" (Armstrong, 1972:225). Vegetational habitat ranges from the lower limits of woodland (N. c. arizonae)
in the Four-corners area to above timberline. This seems to
have been the most wide-spread and common Neotoma during
full-glacial times, having occurred from the highest site ee
(Jaguar Cave) to as low as Dark Canyon Cave and possibly
south to Jimenez Cave.

As with Marmota, much has been made in the literature of
Pleistocene occurrences outside its present range, but generally in comparison with populations in the high forests of
the Southern Rockies of northern New Mexico. Its southern
Pleistocene occurrences probably were in a woodland context.

†Nothrotheriops shastensis. Shasta Ground Sloth.

This ground sloth occurred in all of our zones except the
Northern Highlands. It is known from the Sangamonian of
California and from the mid-Wisconsinan and late Wisconsinan

stadials from Medicine Hat to Dry Cave. This probably is the
taxon represented in the Dry Cave interstadial sites, too.

In the late glacial, it possibly occurred as far north as
Wilson Butte Cave (queried identification), but more certain
records occurred no farther north than the northern Californian caves, Arizona, and New Mexico, thence south into
Mexico.

Notiosorex crawfordi. Desert Shrew.

Notiosorex is the most xeric-adapted of the New World
shrews, occurring from the lower boundary of coniferous forest (Lindeborg, 1960) into the Southwestern deserts.
Although the temptation is to characterize its ecologic requirements as aridity and warmth, its wide ecologic range and
its widespread occurrence in Pleistocene faunas within its
present range suggest that these are tolerances rather than
requirements. A more realistic evaluation may be that such
tolerances allow its existence in areas uninhabitable by
other shrews, thus avoiding their competition. If this is
correct, then occurrence in Pleistocene faunas may imply presence of microhabitats where the cover, food, and moisture
requirements of other shrews could not be met.

Our Pleistocene occurrences were within its present range.
Stadial records were from the mid-range of the Middle-
elevation Savannah Zone into the Steppe Zone (Jimenez Cave).

Ochotona princeps. Pika.

The only pika within our area is an inhabitant of talus
slopes. Limited to high mountain areas, usually near timberline, in the inland and southern portions of our area, it
descends to lowland areas where talus habitat is available in
conjunction with cool and moist conditions. It is absent

from even the most extensive talus deposits in the southern-
most portions of our region. Occurrences at lower elevations
in the Pleistocene implies there was greater effective moisture and possibly cooler temperatures.

Ondatra zibethicus. Muskrat.

Muskrats are widespread and, from our point of view, useful mostly as indicators of permanent water. Thus, for example, stadial occurrence at Baldy Peak Cave in the Florida
Mountains of southern New Mexico strongly suggests that the
Mimbres River was a permanent stream at the time of deposition. It now is normally ephemeral throughout most of its
course from its origin in the mountains more than 80 km to
the north.

Onychomys leucogaster. Northern Grasshopper Mouse.

This mouse occurs from the Prairie Provinces of Canada
south into northern Mexico in the Interior Division, mostly
in upper portions of the Lower Sonoran Life Zone and in Upper
Sonoran habitats (including open woodlands).

Wisconsinan occurrences were within the present range.

Onychomys torridus. Southern Grasshopper Mouse.

The present range is from the southern half of California,
central Nevada, central Arizona, and central New Mexico well
south into Mexico. Reported Wisconsinan occurrences were
within or close to the present general range, but I regard
only the Rancho La Brea record as sure for the time span.
The Isleta Caves are known to have included Holocene material
and the Lower Sloth Cave record seems out of place, judging
from the postulated habitat and the situation at Dry Cave.
Misidentification or contamination by Holocene material is

suspected. The Muskox Cave record could be pre-stadial, but may be a misidentification.

Oreamnos americanus. Mountain Goat.

Presence of this bovid indicates a high boreal to alpine habitat. Specimens listed as Oreamnos sp. may pertain to this species or the following species. Current native distribution is from the southern Yukon south to the high mountains of central Idaho and, in the Cascades, central Oregon. Several of the Wisconsinan occurrences were outside this range (e.g., Bell Cave, Potter Creek Cave).

†Oreamnos harringtoni. Harrington's Mountain Goat.

This mountain goat, known only from presumed stadial deposits, was smaller than O. americanus. Its indicated range was to the south of that species and, in the Interior Division, at lower elevations ee. In terms of stadial vegetative zones, it occurred from the lower portions of the Northern Highlands Zone to the bottom of the Steppe-woodland. It is known only from sites near rugged topography.

†? Oreamnos, undescribed species. Undescribed Mountain Goat.

L. Logan (pers. comm.) has identified what may be a third mountain-goat-like taxon from the Guadalupe Mountains Region of southeastern New Mexico.

Oreortyx pictus. Mountain Quail.

This bird was considered by Johnsgard (1973:345) to be "perhaps the most temperate-adapted of any species" of quail. Leopold (1959:224) reported it in the mountains of northern Baja California as summering "on the high crests, in conifer forests, and around the edge of mountain meadows, dropping

down in winter to the oak and chaparral zone on the western
slope." He also noted that "on both summer and winter
ranges, mountain quail show a decided preference for thick
brush. They never venture far from dense cover, . . ." At
lower, arid elevations it may breed into sagebrush and
pinyon-juniper habitats if free water is available. In most
cases, the presence of brushy habitats seems important.
Historically, it occurred from northern Baja California into
Oregon east into western Idaho and barely into northern and
western Nevada. Significant migratory movements appear to be
limited to birds moving to nest at high altitudes and even in
these cases are on the general order of only some 32 km.

Mountain quail today are even more limited to winter-
dominant precipitation areas than are sage grouse. Past
occurrences would be evidence for brushy or mixed shrub-tree
growth plus the presence of free water within about 1.5 km,
at least during the nesting season.

It occurred in late stadial times as far east as southeastern New Mexico.

†Orthogeomys (Heterogeomys) onerosus. Giant Hispid Gopher.

The sole record of this large pocket gopher was from San
Josecito Cave. Kurten and Anderson (1980) recorded it as
Heterogeomys onerosus, but Russell (1968) placed Heterogeomys
as a subgenus of Orthogeomys. Kurten and Anderson suggested
occurrence of the genus outside its present range may mean
that part of the San Josecito Cave fauna was Sangamonian in
age, but since the species is extinct, such an assumption is
dangerous.

Ovibos moschatus. Muskox.

Muskoxen are tundra animals most closely approaching our

area in extreme northern Manitoba. Their presence in fossil
faunas is considered indicative of tundra. Fossil records
were from Canada and Wyoming.

Ovis canadensis. Mountain Sheep.

Ovis canadensis has been distributed historically through
most of the Interior Division from British Columbia to northern Mexico, and into parts of the West Coast Division. On
the east, populations apparently extended onto the plains.

†Ovis catclawensis, a species based on large size, was
synonymized by Harris and Mundel (1974). Martin and Gilbert
(1978) used this name for the large sheep recovered from
Natural Trap Cave and indicated that not only large size, but
that elongated legs characterized the species. †O. catclawensis is considered a synonym of O. canadensis for our purposes here.

Wisconsinan records were widespread in the Interior Division, but absent from the West Coast Division. All vegetative zones except Steppe were inhabited.

†Paleolama mirifica. Stout-legged Llama.

This is known from a single site in the West, the Emery
Borrow Pit, in California. The fauna was Rancholabrean, but
a more exact age is not published.

†Panthera leo atrox. Lion.

This cat, now considered conspecific with the African
lion, is known from Alaska to Peru (Kurten and Anderson,
1980). In the West, it extended throughout our area and from
our highest site to our lowest. High mountain areas may have
been avoided, but records were well distributed among flat-

land sites and those of foothills and low mountains. Occurrences were from Sangamonian to latest Wisconsinan.

Panthera onca. Jaguar.

Historically, jaguars are recorded north to northern Arizona and New Mexico, though apparently rare at this edge of
their range. Wisconsinan occurrences were earlier than full-
glacial except probably for the southern site of San Josecito
Cave. Smith Creek Cave also recorded jaguar, but several
other taxa indicated that non-stadial elements were recorded
in the faunal list.

Pappogeomys castanops. Chestnut-sided Pocket Gopher.

The current range of this large pocket gopher runs from
southeastern Colorado and southwestern Kansas south well into
Mexico. Russell (1968) hypothesized the retreat of P. castanops from the Southwest during the Wisconsinan pluvial cycle,
with post-Pleistocene re-invasion. Occurrence at several
stadial sites in the Guadalupe Mountains area indicates that
populations were maintained farther north than envisioned by
him. They also were recorded in the Dry Cave interstadial.

Perognathus. Pocket Mice.

A number of species of pocket mice have been recorded from
Sangamonian and Wisconsinan sites, mostly from within or near
their present ranges. Despite the number of species, remains
are relatively scanty in number and many of the pluvial sites
in the Interior Division may have Perognathus as Holocene
contaminants. This is a distinct possibility, for example,
for the two species recorded from the Isleta Caves. Mention
already has been made of the absence of Perognathus in the
full-glacial faunas of Dry Cave.

Peromyscus. White-footed Mice.

This genus is nearly ubiquitous in the West and is represented by many species. Although many of the species are
excellant ecological markers today, species identification is
notoriously difficult--oft times, for even entire, modern museum specimens.

†Peromyscus anyapahensis. Anacapra Mouse.

White (1966) described this large species from West
Anacapra Island, one of the Channel Islands off the coast of
southern California. It is known only from that island.

†Peromyscus cragini. Cragin's Mouse.

This is an extinct white-footed mouse of the subgenus Haplomylomys. It was reported from Mesa De Maya. Other identifications were from Great Plains sites (Kurten and Anderson,
1980).

Peromyscus crinitus. Cliff Mouse.

Cliff mice are closely limited to cliff-type habitats,
from the low desert of the Lower Colorado River to the high
Colorado Plateaus and northwest to north-central Oregon. The
highest elevations occupied support pinyon-juniper growth or,
possibly, marginal ponderosa pine. The area is one of dominant cold-season precipitation, with only the most easterly
populations in an area of an approximately equal cold-warm
season precipitation regime. Its Pleistocene presence to the
southeast of its current range may imply greater winter precipitation; competition with a more easterly species, such as
P. difficilis, cannot be ruled out as a factor in limiting
its expansion to the east today, however.

Tentative identifications have been made of this species

from California and southeastern New Mexico. On ecological grounds, Logan and Black's (1979) identification of P. ere-micus probably referred to this species, since the character used for identification is shared by the two species.

†Peromyscus nesodytes. Santa Rosa Mouse.

This is a large form, limited in occurrence to Santa Rosa Island. White (1966) suggested that the presence of large species of Peromyscus on the Channel Islands may indicate adaptation for a woodrat niche, Neotoma being absent from the islands.

Peromyscus truei. Pinyon Mouse.

This mouse is associated closely with pinyon-juniper wood-land today. The two stadial records are beyond the present range, though that at Blackwater Draw only by a relatively short distance. The latter site also lies between the main modern distribution and the range of an isolated subspecies (or closely related species) to the east, P. t. comanche. The Wilson Butte Cave record was queried in the original work and, since the site is elevationally well above the hypothe-sized woodland vegetative element, seems likely to be incor-rect.

Phenacomys intermedius. Heather Vole.

Heather voles are boreal animals usually associated with high-altitude forests in the West, though high-elevation grass and sagebrush habitats may be occupied. In Colorado today, their elevational range is ca. 6300 to 7850 m ee. The animal is used here as a boreal forest marker.

It occurred in the Sangamonian Silver Creek fauna and in a

number of stadial faunas from Wyoming and Utah north. The
sites were in the Northern Highlands Zone.

Pitymys ochrogaster. Prairie Vole.

Repenning (1983) has presented justification for the use
of Pitymys for the North American voles of this group; most
workers have included them within the genus Microtus.

With a distribution in our area primarily in the plains
from well into Canada south to northeastern New Mexico, the
prairie vole inhabits moderately well to well developed
grasslands. Much of the westernmost distribution is within
the more mesic habitats of major river valleys; the more
scantily-clad interfluves appear to support insufficient
plant growth.

Pleistocene occurrences outside the present range presumably recorded the past presence of more mesic grasslands than
occur there today, perhaps coupled with a summer-dominant
precipitation pattern. In the south, possibly cooler temperature also may have been involved. The Sangamonian Mesa
De Maya site currently is near the edge of this vole's
distribution.

†Platygonus compresssus. Flat-headed Peccary.

These peccaries were present during late-glacial times in
our area in the extreme east (Blackwater Draw, Agate Basin)
and at several sites on the West Coast. It apparently survived into the early Holocene at the Agate Basin Site (Walker, 1982b). The probably-interstadial Fossil Lake site also
included †Platygonus, but as †P. cf. vetus. Skinner (1942)
reported †P. alemani from Papago Springs Cave. This probably
was †P. compressus. Kurten and Anderson (1980) suggested the
animal inhabited open country and that the dentition likely

indicates browsing. Stadial absence in the Interior Division
except in the plains, but occurrence on the West Coast, is a
pattern seen in such forms as †Glossotherium and Procyon
lotor. These forms may well have swung south of our main
area, as indicated for †Platygonus.

†Plecotus tetralophodon. Four-lophed Big-eared Bat.

This bat is similar to the living Townsend's big-eared bat
except for an upper dental feature. It was reported only
from San Josecito Cave, but since lower dentitions are non-
diagnostic, could have been represented elsewhere but reported as Plecotus sp. or P. townsendi.

Procyon lotor. Raccoon.

Raccoons are fairly widely spread in the West today,
though absent from much of the southern Great Basin away from
permanent water. In stadial times, almost all of the Interior Division seems to have lacked the animal. The only records were from the Brown Sand Wedge, where a number of eastern forms apparently reached their western limits, and a
queried Procyon identification from Schuiling Cave. This
virtual absence is puzzling, since they are common fossils to
the east and occurred in several West Coast sites.

Rangifer tarandus. Caribou.

Caribou are arctic deer that now barely reach our territory in northeastern Washington and north-central Idaho.
Pleistocene occurrences south of its present range presumably
would document colder climatic conditions.

It is used here as an ecological marker for boreal forest
and/or as indicative of tundra or near-tundra conditions. It

occurred in stadial sites as far south as southern Idaho
(Shoshone Falls).

Sciurus. Tree Squirrels.
Tree squirrels of any type are rare in the Late Pleistocene record in the West. The most obvious requirement of the
genus is the presence of trees, whether the open ponderosa
pine forest of Abert's squirrel (S. aberti), the riparian
deciduous growth of the Arizona gray squirrel (S. arizonensis), or the varied habitats of the western gray squirrel (S.
griseus). In general, these squirrels utilize tree products
(vegetative parts, mast, etc.).

Sciurus arizonensis. Arizona Gray Squirrel.
Slaughter (1975) reported S. cf. arizonensis from the
Brown Sand Wedge fauna of Blackwater Draw. As I have commented elsewhere (Harris, 1970b), this makes no sense biogeo-
graphically, but S. carolinensis (apparently not considered
by Slaughter) would. If S. carolinensis is the correct species, it would be one of several eastern forms reaching its
western limits along a riparian stringer of the Brazos drainage.

Sigmodon. Cotton Rats.
Sigmodon is a useful ecological indicator in that it
appears to be directly affected by climatic parameters.
Mohlhenrich (1961:22) showed that "in New Mexico, cotton rats
generally occur only in areas which have a mean annual temperature of 55.0°F or above, a mean January temperature above
34.0°F, and a mean July temperature of 75°F or above." He
also pointed out that the growing season generally is at
least 180 days long. Although mean annual temperature might

not be meaningful under some postulated Pleistocene temperature regimes, the maximum and minimum mean temperature parameters should have prevailed.

Although present in the Dry Cave interstade, the easily
recognized Sigmodon was entirely absent from the well-
documented New Mexican stadial faunas with the exception of
the Brown Sand Wedge and the Dry Cave sites TT II and Human
Corridor. Presence at these late sites may have heralded
warming tempertures at the end of the Pleistocene. Sigmodon
also appeared in the relatively low-elevation Murray Springs
Arroyo fauna in Arizona and at Jimenez Cave in southern Chihuahua.

†Smilodon fatalis. Sabertooth.

†Smilodon was limited to the Gray Sand fauna of Blackwater
Draw within the stadial, Interior Division sites. The Cady
local fauna had a queried occurrence and several West Coast
sites had this cat as a faunal component. The range may have
been continuous to the south, a record being available from
San Josecito Cave.

Sabertooths were more widespread in the interglacial and
interstadial sites, ranging from Canada (Medicine Hat) to
Idaho and Utah, as well as having occurred on the West Coast.

Sorex cinereus. Masked Shrew.

This shrew showed major changes in geographic distribution
during the Late Pleistocene. Although generally thought of
as a boreal, high mountain forest species in the West today,
in fact it may be more of a cool-climate hydrosere inhabitant
and restricted to highland areas in the south because only in
such highlands are its temperature requirements met. Its
Pleistocene presence, then, may indicate areas of cool sum-

mers with abundant low vegetation, such as in forest clearings near permanent water or, in lowland areas, sedge beds or marshy grassland in riparian situations. Its absence today from seemingly suitable lowland areas throughout most of our area presumably is due more to the stressful high temperatures than to unsuitable vegetation. Likely its temperature restrictions are more severe than those of Zapus and Microtus pennsylvanicus, which to some degree share this pattern, for it lacks the Southwestern relictual populations of these taxa.

Sorex hoyi. Pygmy Shrew.

This is a trans-boreal species moving south in the Northern Rockies to western Montana and with isolated populations in southern Wyoming to central Colorado. It is taken as an indicator of high-boreal conditions. The only fossil occurrence was east of its present range, at Little Box Elder Cave.

Sorex merriami. Merriam's Shrew.

Although overlapping shrews of the monticolus/vagrans group ecologically (Jones, 1961), Merriam's shrew appears to be adapted to yet more arid and probably slightly warmer conditions. Although mostly known from the more xeric coniferous forests of the West, it recently has been reported from pinyon-juniper woodland in west-central New Mexico (Diersing, 1979). As with the other Sorex, food, cover, and critical climatic conditions seem more important than the type of vegetation.

As a fossil, it is known only from the sites of southeastern New Mexico, but was common in the Dry Cave stadial sites.

Sorex monticolus/vagrans. Vagrant-like Shrews.

Shrews long recognized as S. vagrans recently have been considered to consist of two species (Hennings and Hoffman, 1977)--most literature records these as S. vagrans, and separation is infeasible on the basis of published paleontologic descriptions. Current distribution generally is with S. vagrans in much of the Great Basin, replaced by S. monticolus in much of the Southwest.

These are shrews of mixed coniferous forest, generally of drier conditions and warmer temperatures than common with S. cinereus or S. palustris, though the moister habitats within the general vegetational type are preferred. Where the range of S. cinereus overlaps, that species tends to exclude these shrews from the wetter habitats. As with the masked shrew, likely availability of prey animals and general moisture and temperature conditions are limiting rather than vegetation type per se. Thus presence in Pleistocene deposits may not necessarily indicate mixed coniferous forest, but rather selected climatic parameters that today are associated with that vegetational type.

Both Findley (1965) and Harris (1970b) noted that their samples (Hermit's Cave and stadial Dry Cave, respectively) were of smaller individuals than currently is the case with the nearest living population. They thus more closely resembled more northern populations. Fossil records are stadial, in southern New Mexico (Baldy Peak Cave east to the Guadalupe Mountains) and adjacent Texas.

Sorex nanus. Dwarf Shrew.

Too few modern specimens of this small shrew are available to allow more than a general characterization of its requirements. It would seem to best fit into the ecological series

noted below at about the same range as the vagrans-like
group, though possibly with a somewhat more mesic and cool
aspect.

It occurred at Hermit's Cave, in the Guadalupe Mountains,
and probably is represented by a single specimen from the Dry
Cave full stadial (Animal Fair).

Sorex palustris. Water Shrew.

The common name of this shrew is a giveaway as to its habitat, for it is found exclusively around the margins of permanent water; as it is not known from permanent water sources
at lower elevations in the southern portions of its general
range, high summer temperatures also likely limit its distribution today. Most Pleistocene records were within its present general distribution, though at lower elevations. However, Logan (1981) has identified this species at Muskox
Cave, far to the south of its current range.

The Interior Division long-tailed shrews (genus Sorex)
seem to form a roughly graded series (Sorex hoyi, S. palustris, S. cinereus, S. nanus, S. monticolus/vagrans, S. merriami) from habitats characterized by cold winter temperatures,
cool summer conditions, and plentiful moisture at one end of
the series to relatively xeric, though well vegetated, habitats with moderately cold winters and moderately warm summers
at the other end. All would seem to require sufficient
moisture spread throughout the year to support both adequate
cover and abundant invertebrate life for food.

Spermophilus. Ground Squirrels.

Although of potential use in paleoecology, separation of
species on incomplete material is difficult. In some cases
(e.g., S. spilosoma and S. tridecemlineatus), slight size

differences have been used to separate forms osteologically similar, but geographic variation among living populations may exceed the size criteria used.

Spermophilus lateralis. Golden-mantled Ground Squirrel.

Inhabitants of open timber or meadows edged by timber (Hall and Kelson, 1959), this squirrel is widespread throughout the montane West. It is, however, absent from mountain ranges south of the Mogollon Rim of Arizona and from mountain ranges that are both east of the Rio Grande and south of the Sangre de Cristo Mountains in New Mexico. Pleistocene occurrences outside this range presumably imply at least greater effective moisture, but other implications are unclear.

Its occurrence at Ventana Cave is of particular interest. Considering the pattern seen in the Interior Division, long-range transport seems probable, despite the statistical unlikelihood. It has not been identified from the southern New Mexican ranges east of the Rio Grande despite a good fossil record, indicating that southward movement in the Rockies or eastward movement to northern New Mexico probably occurred after these ranges were separated from the Sangre de Cristo Mountains by xeric conditions.

Spermophilus richardsoni. Richardson's Ground Squirrel.

This species is of interest not only because it has been identified well outside its current range during the Pleistocene, but also because historical and archaeological records indicate large range fluctuations during late Holocene times.

This squirrel currently occupies cool grasslands from well north of the U.S. border to central Colorado and Nevada, but avoids most of the plains area south of northeastern South Dakota and central Montana; most populations in our area are

montane. Archaeological records from Arroyo Hondo Pueblo
near Santa Fe, New Mexico, indicated occurrence far south of
its present range as recently as 800 to 1000 years ago (Lang
and Harris, in press). Until such fluctuations are understood, too much reliance on this taxon as a paleoecologic
indicator is dangerous.

A number of Pleistocene sites in the central and northern
parts of the Interior Division recorded this species. A
queried identification based on a single tooth is available
from a Dry Cave site.

Spermophilus variegatus. Rock Squirrel.

This large ground squirrel is associated with rocky areas
from desert into the Transition Life Zone. The Little Box
Elder Cave record was north of its present range and probably
was Holocene.

Although recorded in a number of Guadalupe Mountains
sites, the stadial situation in New Mexico is not clear.
Despite the tremendous number of stadial fossils examined
from Dry Cave, the rock squirrel has not been recognized (it
was present in the interstadial Room of the Vanishing Floor,
however). It occurs in the area today and even with a greater soil mantle and with dense vegetation in drainageways,
habitat should have been available if the animal had not been
excluded from the region. Likewise, it was unidentified in
the extensive collections from the Isleta Caves.

Spilogale putorius/gracilis. Spotted Skunks.

Spotted skunks of the U.S. have alternately been considered to consist of two species, the western S. gracilis and the
eastern S. putorius, or of a single species, S. putorius.
With little value paleoecologically, all specimens given to

species in the literature are listed here as S. putorius,
with no implications as to the validity of one versus two
taxa intended. The Medicine Hat 5 specimen (identified to
the "cf." level) was far north of the current range, which
extends north at nearest to extreme southwestern Montana.

†Stockoceros conklingi. Conkling's Pronghorn.
    Records are from the Interior Division stadial sites from
the bottom of the Middle-elevation Savannah to our lowest
site in the Steppe Zone. All occurrences were from foothills
or low mountains in Arizona, New Mexico, and northern Mexico.

†Stockoceros onusrosagris. Quentin's Pronghorn.
    The known elevational range of this pronghorn is more re-
stricted than that of †S. conklingi (the possibility of con-
specificity is noted by Kurten and Anderson, 1980, however).
Logan (1981) recorded both from Muskox Cave (with this species assigned to †Tetrameryx). The lowest site from which it
was recorded is Burnet Cave, only 114 m ee lower. Sites were
limited to southern Arizona and New Mexico.

Sylvilagus. Cottontails.
    Three species of considerable paleoecologic import occur
in the Interior Division: S. auduboni, S. floridanus, and S.
nuttalli. Unfortunately, although identification to the family Leporidae is easy, as is, with most of our taxa, separation into hares and cottontails, identification to species
presents great difficulties.
    The most useful features for identification of cottontail
species are enamel pattern of lower cheekteeth, depth of dentary relative to toothrow length, basioccipital constriction,
and relative bullar size.

The depth of dentary separates most $\underline{S}$. auduboni from the other two species (Findley et al., 1975--but note that their Fig. 35 is mislabeled and should read alveolar length of P_3-M_3). The desert cottontail has a deeper dentary than do either of the other two species. An adult or near-adult P_3-M_3 length of <13.8 mm and an index (dentary depth at P_4 divided by P_3-M_3 length, times 100) of 84 or more usually indicates $\underline{S}$. auduboni, although occasional individuals will be misidentified.

$\underline{Sylvilagus}$ $\underline{nuttalli}$ usually can be told from $\underline{S}$. floridanus and $\underline{S}$. auduboni by the amount of enamel crinkling in the lower cheekteeth. Not only P_3, but also subsequent lower teeth have little or no crinkling, while the other species (particularly $\underline{S}$. auduboni) normally show such crinkling. The character of greater crinkling appears somewhat more constant in the post-P_3 teeth of $\underline{S}$. floridanus than in the P_3.

The basioccipital constriction is much greater in $\underline{S}$. auduboni than in the other two species. This is measured between the most medial projections of the two auditory complexes. Only in obviously very young individuals of the other two species (n=44) is this width less than 6.1 mm, while the maximum in a sample of $\underline{S}$. auduboni (n=24) is 6.0 mm. An index between the basioccipital constriction and the auditory bullar width (measured from the most medial auditory projection to the lateral surface ventral to the auditory tube of the same side) separates even very young individuals. Of 24 $\underline{S}$. auduboni, 22 $\underline{S}$. floridanus, and 22 $\underline{S}$. nuttalli, the largest index for $\underline{S}$. auduboni is 59.2; the smallest for $\underline{S}$. floridanus, 63.2; and, with one exception, the smallest for $\underline{S}$. nuttalli is 65.7. The exception is from an area of sympatry between $\underline{S}$. auduboni and $\underline{S}$. nuttalli where possible hybridiza-

tion is occurring (Findley et al., 1975). This individual
has an index of 57.1.

Sylvilagus auduboni. Desert Cottontail.

This is the most widely distributed cottontail in the
West, occurring throughout the lower elevations of the southern half and, in the eastern part of our area, north to
northern Montana. The common name implies a much greater
restriction than actually occurs, for it is also common in
the grasslands of the Southwest and enters the pinyon-juniper
woodlands and sagebrush stands.

The distribution of cottontails strongly suggests competitive exclusion is common. True sympatry between species is
rare. Ordinarily, overlap zones are narrow and even within
such zones, different microhabitats tend to be inhabited
(Harris, 1963a). In contact zones, S. auduboni tends to inhabit the opener, more xeric areas, replaced in heavier, more
mesic habitat by whichever of two species, S. nuttalli or S.
floridanus, is available. Thus replacement of S. auduboni by
one of these species implies lusher, thicker vegetation.

Sylvilagus floridanus. Eastern Cottontail.

To the east, S. auduboni typically is replaced by S. floridanus in taller grasslands and riparian situations. In much
of the Southwest, however, altitudinal changes in vegetation
from Upper Sonoran grasslands and poorly developed woodlands
to better developed woodlands and coniferous forests also are
matched by a switchover to the eastern cottontail. Often
chaparral-like growth is favored by the latter. Pleistocene
occurrences of the eastern cottontail in areas now inhabited
by the desert cottontail thus imply the presence of lusher
grassland or more densely vegetated brushlands; which vegeta-

tive type is involved probably depends upon the topography in part, grasses usually having the advantage on deeper-soiled, more level terrain.

The present distribution of S. floridanus compared to that of S. nuttalli suggests the former has the advantage in areas where warm-season precipitation is dominant. The overall impression is that S. floridanus basically is a grassland form, abandoning such habitat only when forced into islands of montane vegetation by competition with S. auduboni caused by increasing aridity of the surrounding grasslands.

Occurrence of S. floridanus at the Isleta Caves may be hypothesized as later than the occurrence of S. nuttalli, heralding the arrival of late Wisconsinan-early Holocene warming and possibly a switch-over to more warm-season precipitation.

†Sylvilagus leonensis.   Southern Pygmy Rabbit.

This small rabbit was described from San Josecito Cave in Nuevo Leon.

Sylvilagus nuttalli.   Nuttall's Cottontail.

Nuttall's cottontail replaces S. auduboni on the north and S. floridanus on the west and in the northern highlands of New Mexico and Arizona. In the northern portions of our region, Nuttall's cottontail spreads into the extensive sagebrush areas; farther south, montane habitats are inhabited almost exclusively. In much of its range, this rabbit occurs in areas with much of the precipitation occurring during the cold season. In comparison with S. floridanus, S. nuttalli seems basically a shrubland form rather than a grassland type, but also extends into montane habitats where it is free from competition from other cottontails.

†<u>Symbos</u> <u>cavifrons</u>. Woodland Muskox.

Within our basic suite of sites, †<u>Symbos</u> occurred only in the Northern Highlands Zone in late-glacial times. However, occurrence at Black Rocks in western New Mexico probably was stadial. This site, at 5765 m ee, would fall into the Middle-elevation Savannah Zone. It also was recorded from the Alpine Formation near Great Salt Lake in Utah at a similar elevation, though the formation generally has been considered early or middle Pleistocene. Geist (pers. comm.) suggests it too is late Wisconsinan.

†<u>Tadarida</u> <u>constantinei</u>. Constantine's Free-tailed Bat.

This extinct bat was about 10% larger than the common Brazilian free-tailed (<u>T</u>. <u>brasiliensis</u>) and showed some proportional differences from that species (Lawrence, 1960). It is known only from deposits (>17,800 BP) in New Cave, in the Guadalupe Mountains region of New Mexico.

<u>Tamias</u>. Chipmunks.

A number of species of chipmunks occur in the West. In the southern portions, one species or another occurs from the woodlands on up through the highest habitats. Farther north, sagebrush and semi-desert scrub habitats as well as higher vegetative types up into the alpine zone are inhabited (<u>Tamias</u> <u>minimus</u>).

Most fossils are identified to genus only. Those in the south may be taken as probably indicating at least the presence of woodlands, if not higher-elevation communities.

<u>Tamias</u> <u>cinereicollis</u>. Gray-necked Chipmunk.

<u>Tamias</u> ? <u>cinereicollis</u> was recovered (UTEP) from a presumed late Wisconsinan stadial site (Baldy Peak Cave) in the

Florida Mountains of southern New Mexico, which would imply
that woodland was continuous across the Deming Plains to the
north. Woodland elements occur in the range now, although
chipmunks are not recorded, but the Plains are a notable barrier to woodland and higher-zone mammals.

Tamias dorsalis. Cliff Chipmunk.
    Tamias ? dorsalis was recorded from Papago Springs Cave.
This species generally is an oak or pinyon-juniper woodland
form, occurring slightly to the northeast of the area today.

Tamias minimus. Least Chipmunk.
    This is the species most commonly identified as a fossil
to the specific level. Its wide ecological range renders it
of little aid in ecological reconstruction.

Tamias umbrinus. Uintah Chipmunk.
    Tamias cf. umbrinus was recorded from Smith Creek Cave, at
the margin of its occurrence today.

Tamiasciurus hudsonicus. Red Squirrel.
    The red squirrel is a tree squirrel and a particularly
good ecological marker in the West since it generally occurs
only in mountains supporting spruce-fir forest; in such mountain ranges, however, it may enter into mixed coniferous
forest on occasion.
    It was identified only from the Guadalupe Mountains area
of New Mexico, where it indicates a spruce-fir forest connec-
tion between the southern New Mexican mountains and those to
the north. This implies descent of such forest to lower than
5600 m ee in central New Mexico. Stearns (1942) recorded a

squirrel the size of the red squirrel from New La Bajada Hill
in northern New Mexico at an elevation of ca. 5600 m ee.

†Tapirus. Tapirs.

Tapirs now approach the U.S. to no nearer than southern
Mexico. Presumably-extinct species were recorded from several sites in the West Coast Division (Sangamonian and later)
and from stadial sites in Arizona. In the latter sites, presumably Sagebrush Steppe-woodland to Steppe was inhabited. A
record also is available from the interstadial Lost Valley
fauna of Dry Cave.

Tayassu tajacu. Collared Peccary.

This wild pig currently occurs in the USA from central
Texas into the eastern Trans-Pecos and from southwestern New
Mexico west and north into central Arizona. From these
areas, it continues south into South America.

The only record from our area that is outside its present
range is from Dark Canyon Cave, where a canine in the Texas
Memorial Museum collection appears to represent this species.

Thomomys bottae. Botta's Pocket Gopher.

This pocket gopher is widespread in the West from southwestern Oregon southeasterly to northeastern New Mexico. Its
gross range overlaps that of T. talpoides in the north. In
this area of overlap, T. bottae generally inhabits the lower
elevations and T. talpoides the higher, montane habitat.
Where T. bottae overlaps with Geomys and/or Pappogeomys, it
abandons the deeper valley and bolson soils in favor of montane habitats.

Distribution in Interior Division stadial sites was in the
south only (Isleta Cave No. 1, where it may have been Holo-

cene, and southern New Mexico) and from the middle of the Middle-elevation Savannah to the bottom of the Sagebrush Steppe-woodlands.

Thomomys talpoides. Northern Pocket Gopher.

Thomomys talpoides extends from well into Canada south to the mountains of northern Arizona and New Mexico. North of the range of T. bottae, it occupies both lowland and mountains. In the south, populations are isolated on islands of high mountain vegetation, separated from other enclaves by the encircling T. bottae.

Luckily, the two species are fairly reliably separated by both skull and mandible characters. The nature and position of the infraorbital canal relative to the incisive foramen differ between the two (more anterior, with lateral wall of opening forming a larger arc and medial wall less narrowly depressed in T. bottae). In the dentary, the trigonid of the lower fourth premolar of T. bottae almost always "is bilaterally symmetrical while that of T. talpoides is not" (Miller, 1976:396). Miller also pointed out that the first and second lower molars "in T. bottae have pronounced labial sulci, while these molars in T. talpoides usually do not." Also, the masseteric shield on the lateral surface is more acute at the anterior end in T. talpoides than in T. bottae.

Presence of T. talpoides at low to moderate elevations in Pleistocene faunas outside its present range implied both cooler average temperatures and more effective moisture than now. The two species occurred together in several Dry Cave stadial sites, suggesting that T. bottae inhabited the relatively warm, dry south-facing slopes and T. talpoides the cooler, more mesic north-facing slopes.

Urocyon cinereoargenteus. Gray Fox.

Widespread in eastern USA, the gray fox is absent from the West north of the latitude of Wyoming except for an extension up the West Coast to northern Oregon. It generally inhabits woodland vegetation, but does occur into ponderosa pine forest and, in rough topography, down somewhat into Lower Sonoran situations.

Despite listing in several sites, Urocyon appears to have been rare in the stadial Interior Division. There was possibly one specimen from the Dry Cave stadial sites (but occurrence in all three interstadial sites despite the much smaller number of bones recovered), and a single record from Isleta Cave No. 2 (possibly Holocene). Williams Cave included Holocene material, and not all of the sites within Anthony Cave were surely stadial. Only in the western portion of the Interior Division does presence during stadial time seem relatively sure.

Vulpes macrotis. Kit Fox.

Vulpes macrotis and V. velox are considered to be two separate species, although a recent trend has been to consider them conspecific (Kurten and Anderson, 1980; Hall, 1981).

This is an animal of the desert and western, arid grasslands. Occurrence at the Isleta Caves almost certainly was Holocene. It was reported tentatively from Jaguar Cave (currently outside its geographic range), but V. velox seems more likely. During stadial times, this animal appears to have been limited to the West Coast and the western portion of the Interior Division, probably in the Steppe Zone.

Vulpes velox. Swift Fox.

This is a more eastern form than V. macrotis, extending in

grasslands from the Prairie Provinces of Canada to extreme
southeastern New Mexico and adjacent Texas today.

The Pleistocene form was notably larger than the living
form, and fossils from New Mexico currently are under study
by Becky Provost (UTEP). All the New Mexican sites except
Blackwater Draw are in the current range of V. macrotis.
Thus V. velox moved westward during the late glacial, replac-
ing the kit fox (the record at Smith Creek Cave was listed as
V. velox, but it is unknown whether the two-species taxonomic
concept was employed).

Vulpes vulpes. Red Fox.

This larger fox tends to be a forest animal, though known
from grassland context on occasion. Hall (1981) speculated
that recent movements of red foxes westward in Kansas may
have been allowed by a decrease in the coyote population. If
so, differences in competition might change ecological "pref-
erences" of this animal significantly.

Several stadial sites were beyond the current range of the
red fox. The Isleta Caves records probably were Pleistocene.
The Burnet Cave identification probably is correct, but the
Brown Sand Wedge record might pertain to V. velox. Slaughter
(1975:184) referred an upper first molar (in his 1964 pre-
print of this paper, it was the first lower molar--the latter
almost certainly is correct) to V. vulpes, stating that "the
New Mexico specimen is too large for the Swift Fox and larger
than our specimens of Kit Fox." Since he presumably was not
aware of the larger size of Wisconsinan-age V. velox, that
species may have been involved.

Several species are undescribed taxonomically, but are
recorded in the appendices. Brief notice of these follows.

†Leporidae, undescribed rabbit.

The interstadial sites at Dry Cave contained remains of a very small rabbit apparently not ascribable to any extant genus. This taxon currently is under study by B. Russell and myself.

†Neotoma, Species A, undescribed woodrat.

This species is somewhat intermediate in its observed characters between N. cinerea and N. mexicana, but probably is closest to the former. It was recorded only from interstadial sites at Dry Cave and may possibly represent a morphologically divergent population isolated after an earlier expansion of N. cinerea (Harris, in press, a, b).

†Neotoma, Species B, undescribed woodrat.

This is a very small-sized Neotoma recorded from the Dry Cave interstadial sites. It seems to be related most closely to N. goldmani, though differing significantly from that taxon in a number of characters (Harris, in press, a, b).

APPENDIX 2. DATA BASE TAXA

Appendix 2 lists each taxon recorded from sites judged to be
of Sangamonian or Wisconsinan age. Presentation is alphabetical except for undescribed species, which are grouped at
the end of the appendix.

The appendix has several drawbacks, not the least of which
is incompleteness; undoubtedly, errors of commission also are
present. One problem inviting inaccuracy is that of changing
nomenclature through the years. I have attempted to apply
currently accepted synonyms to names judged to have undergone
nomenclatural changes, but such a procedure is dangerous to
all except (possibly) the most dedicated nomenclaturalist (to
which I plead innocence).

Another source of error in Appendix 2 lies in site judge-
ments. Some sites, particularly of single finds, may have
been judged as within our time span while actually they are
not; the reverse likely is true also.

Numbers to the left of the taxa are used for identification of the taxa in Appendix 3. Numbers following the taxa
are the numbers of the sites listed in Appendix 3. Identifications less than certain are indicated by a "cf" or "?" before the site number (identifications originally modified by
"cf.," "ref.," or "nr." have been interpreted as "cf"). Extinct taxa are marked by †.

1 †<u>Acinonyx</u> <u>trumani</u>   87, 176, 216, 218

2 Aegolius funereus   279

3 Alces alces   ? 175, ? 196, 300

4 Ammospermophilus leucurus   cf 162, 286

5 Ammospermophilus nelsoni   cf 190

6 Antilocapridae   36, 185, 218, 246, 260, 281

7 Antilocapra sp.   ? 6, ? 52, 82, ? 86, 108, ? 137

8 Antilocapra americana   4, 26, 31, cf 46, 72, 79, cf
   123, 142, 143, 153, cf 154, 155, 157, 162, 167, 175,
   180, 187, 190, ? 193, ? 194, ? 196, 202, cf 216, 218,
   242, cf 248, 275, cf 284, 286, 328, 345

9 Antrozous pallidus   91, 107, 153, 157, 182, 190, 216,
   223, 228, 240, ? 286, 324

10 Aplodontia rufa   138, 240, 259

11 †Arctodus sp.   cf 46, 52, 153, 154, 160

12 †Arctodus simus   6, 83, 120, 123, 175, 190, 218, 240,
    242, 246, 248

13 Baiomys sp.   14

14 Bassariscus sp.   9, 22, 82, 257

15 Bassariscus astutus   cf 10, 46, 162, cf 182, 216,

240, 247, 248, 286, 324, 328

16 †Bassariscus sonoitensis   228, 261

17 Bison sp.   8, ? 38, 50, 57, ? 58, 59, 68, 75, 76, 82, 83, 99, 137, 138, ? 139, 155, 160, 163, 164, 169, 173, 189, 191, 194, 199, 211, 212, 213, 214, 215, 217, 240, 264, 268, ? 286, 294, 300, 313, 314, 316, ? 326, 327, 337, 342, 347

18 Bison or Bos   143, 153

19 †Bison alaskensis   2, 6, 49, 84, 109, 180, 188, 230, cf 336

20 †Bison alleni   cf 12, cf 180

21 †Bison antiquus   4, 11, 13, 17, 20, 25, cf 31, cf 35, 36, 41, 43, 45, 46, 48, 52, 60, 70, 73, 77, 81, 85, 90, 95, 97, 104, 108, 109, 112, 118, 130, 168, 172, 177, cf 180, 188, 190, 201, 228, 246, 248, 258, 260, 266, cf 273, cf 312, 318, 321, 331, 335, 341, 351

22 Bison bison   26, 72, 79, 167, 175, 218, cf 221, 242, 347

23 †Bison latifrons   2, 6, 85, 120, cf 196, 203, cf 222, 229, 246, 248, 256, 260, cf 262, 271, ? 284

24 Brachylagus idahoensis   125, 126, 154, 155, ? 162, 246, cf 284, 286, ? 320, 335, 347

25 †Camelidae   74, 87, 90, 100, 113, 127, 145, 185, 189, ?
   191, 228, 261, 272

26 †Camelops sp.   4, 8, cf 16, 36, 46, 58, 59, 60, 70, 79,
   82, 99, cf 106, 114, 115, 117, 135, 160, 162, 171, 172,
   180, 186, 192, 211, 212, 213, 214, 215, cf 216, 226, 234,
   255, cf 258, 263, 264, 266, 268, 279, 281, 286, 294, 342,
   343, 347

27 †Camelops hesternus   cf 6, cf 26, cf 48, ? 50, 52, ? 58,
   cf 85, 108, 109, cf 112, cf 120, 123, 126, 134, cf 143, cf
   149, cf 153, cf 155, cf 163, 167, cf 175, 177, cf 181,
   190, 193, 194, cf 195, 196, 199, 206, cf 218, cf 242, cf
   246, 248, 250, cf 257, 260, cf 262, ? 270, 273, cf 284,
   293, 295, 318, 319, 320, cf 325, cf 328

28 †Camelops huerfanensis   cf 6, 95, 148

29 †Camelops minidokae   cf. 6, aff ? 52, cf 185

30 Canis sp.   4, 107 (large), 113, 120, 123 (smaller than
    coyote), 139, 216 (wolf), 275

31 †Canis dirus   6, 13, 52, cf 58, 61, 82, 84, cf 85, cf
   91, cf 123, cf 138, 141, 155, cf 162, cf 163, cf 176,
   177, 180, 187, 190, cf 193, 196, 210, 216, 218, cf 222,
   240, 246, 248, 259, 260, 261, cf 262, cf 284, 301, 327,
   342, 344, cf 345

32 †Canis dirus or lupus   135, 182

33 Canis familiaris   155, cf 248

34 Canis latrans   3, 4, 6, 9, cf 10, 26, 36, 46, ? 52, 58,
    72, 82, cf 85, cf 86, 91, 99, 123, 137, 138, 153, 154,
    155, 157, 162, cf 163, 175, cf 180, 181, 187, 190, cf
    194, 199, 202, 209, 210, 218, 228, 246, 248, cf 255, cf
    258, 259, 261, 279, ? 284, cf 286, 301, 307, 320, 322,
    327

35 Canis lupus   4, 9, 26, 27, 35, 36, 46, 72, 91, cf 123,
    141, cf 153, 154, cf 155, 164, 175, 187, 190, 191, 192,
    196, 218, 228, 236, 240, 242, 248, 255, 257, 259, 261, cf
    266, cf 270, 286, 327

36 †Capromeryx sp.   9, 35, 82, 85, 91, 98, 149, 157, 180,
    181, 236, 255, 257, 279, ? 286

37 †Capromeryx minor   34, 190, 248, cf 260, 262, 270

38 Castor sp.   6

39 Castor canadensis   cf 123, 127, 155, 156, 175, 246,
    259, 294

40 Centrocercus sp.   4, ? 162, cf 284

41 Centrocercus urophasianus   82, 123, 145, cf 153, cf
    154, 155, 246, 279, 286, 294

42 †Cervalces scotti   120

43 Cervidae   85, 218, 246, 327 (large)

44 Cervus sp.   190, 228

45 <u>Cervus</u> <u>elaphus</u>   26, 77, 155, 175, cf 196, 201, 268, 286, 333, cf 345

46 <u>Clethrionomys</u> <u>gapperi</u>   4, cf 175, 176, 347

47 <u>Coendu</u> sp.   ? 261

48 <u>Conepatus</u> sp.   216

49 <u>Conepatus</u> <u>mesoleucus</u>   cf 46, 50, 236, cf 261

50 <u>Cryptotis</u> sp.   175

51 <u>Cryptotis</u> <u>mexicana</u>   261

52 <u>Cryptotis</u> <u>parva</u>   9, 107, 145, 157, 182, 216, 312, 324

53 <u>Cuon</u> <u>alpinus</u>   261

54 <u>Cynomys</u> sp.   10, 72, 91, 157, 175, 255, cf 258

55 <u>Cyonomys</u> <u>gunnisoni</u>   cf 26, 153, 154, cf 182, 199, cf 345

56 <u>Cynomys</u> <u>leucurus</u>   cf 143, cf 246

57 <u>Cynomys</u> <u>ludovicianus</u>   35, 46, cf 82, 108, 118, 136, 145, 181, cf 194, cf 196, 199, cf 236, 327

58 <u>Cynomys</u> (<u>Leucocrossuromys</u>) sp.   137

59 †<u>Dasypus</u> <u>bellus</u>   35

60 †<u>Desmodus</u> <u>stocki</u>   75, 240, 261

61 <u>Dicrostonyx</u> <u>torquatus</u>   26, cf 155, cf 175, 176, 218, 242

62 <u>Didelphis</u> <u>virginianus</u>   35

63 <u>Dipodomys</u> sp.   58, 82, 85, 145, 157, 162, 181, 210, cf 233, 255, 257, ? 270, 320, 338, 348, 349, 350

64 <u>Dipodomys</u> <u>agilis</u>   cf 223, 248

65 <u>Dipodomys</u> <u>ingens</u>   cf 190

66 <u>Dipodomys</u> <u>ordi</u>   46, cf 149, 153, cf 286, cf 335

67 <u>Dipodomys</u> <u>ordi</u> or <u>D</u>. <u>merriami</u>   154

68 <u>Dipodomys</u> <u>spectabilis</u>   9, cf 31, cf 91, 145, 149, 153, 154, 181, 312

69 †<u>Edentata</u>   90, 114, 189, 255

70 †<u>Elephantidae</u>   90, 136

71 <u>Eptesicus</u> <u>fuscus</u>   26, 31, 107, 137, 145, 175, 182, 216, cf 261, cf 294, 312, 324

72 †<u>Equus</u> sp.   7, 10, 16, 38, 42, 53, 54, 66, 68, 70, 71, 74, 75, 82, 83, 87, 90, 103, 111, 113, 114, 115, 117, 127, 140, 141, 142, 144, 145, 150, 157, 158, 160, 163, 169, 171, 173, 180, 181, 187, 189, 202, 206, 211, 212,

213, 214, 215, 216, 217, 223, 224, 226, 227, 236, 242,
246, 247, 251, 255, 260, 262, 263, 264, 269, 272, 273,
280, 281, 291, 293, 304, 313, 316, 323, cf 326, 340, 342,
343, 347

73 †Equus sp. (large)   26, 52, 108, 143, 155, 175, 185,
199, 270, 279, 284, 286, 318, 320

74 †Equus sp. (small)   cf 85, 108, 123, 135, 185, 270, 279,
286, 289, 318, 337

75 Equus caballus   ? 26, cf 320

76 †Equus conversidens   9, 26, 30, cf 31, 36, 46, 48, ? 52,
77, 79, 91, 109, 112, cf 120, 121, cf 136, 137, cf 139,
149, cf 154, 155, cf 162, 166, 167, cf 175, 191, 192,
193, 194, 195, 196, 199, 218, 228, 230, 236, cf 258, 261,
? 284, 312, 345

77 †Equus excelsus   cf 6, cf 8, 266

78 †Equus fraternus   ? 65

79 †Equus giganteus   cf 194

80 †Equus laurentius   cf. 6

81 †Equus niobrarensis   9, 31, 35, 36, 46, 50, 91, 137, cf
153, cf 154, 236, cf 258, 312, cf 336, 337

82 †Equus occidentalis   cf 9, cf 58, cf 85, cf 135, ? 138,

? 149, cf 164, 180, 190, 240, 248, cf 255, 259, cf 266, 267, 327

83 †Equus pacificus   123, cf 164, 240

84 †Equus scotti   cf 6, 112, 120, 196, 268

85 †Equus tau   cf 12, cf 46, 228

86 †Equus (Amerhippus)   196, 218

87 †Equus (Asinus)   57, cf 320

88 †Equus (Hemionus)   218

89 Erethizon sp.   348

90 Erethizon dorsatum   9, 26, 72, 107, 137, 139, 153, 155, 175, 182, 196, 216, 247, 259, ? 284, 286, 328, 345

91 †Euceratherium sp.   1, 142, 202

92 †Euceratherium collinum   46, 138, 161, 162, 190, 216, 240, 259, 261

93 Felidae   281

94 Felis or Lynx   318

95 Felis sp.   123 (small), 213, 240, 244, 257, 273

96 Felis (Puma)   52, 320

97 Felis concolor   9, 46, 72, 82, cf 85, 138, 139, 155,
    157, 175, cf 181, 216, cf 240, 247, 248, cf 259, 261, cf
    262, cf 270, 286, 324, 345

98 Felis yagouaroundi   261

99 Geococcyx californianus   58, 82, 91, 157, 190, 248,
    261, 279

100 Geomyidae   181, ? 279

101 Geomys sp.   36, 82, 136

102 Geomys bursarius   cf 35, 108, 199

103 Glaucomys sabrinus   240, 259

104 †Glossotherium sp.   52, 316

105 †Glossotherium harlani   6, cf 85, 108, cf 123, 138,
    180, 190, 246, 248, 251, cf 260, 273, cf 284, 332

106 Gulo gulo   26, 72, 105, 155, 175, 218

107 †Hemiauchenia sp.   9, 65, 82, 85, 123, 135, 149, 162, ?
    175, 180, 196, 202, 223, 248, 270, ? 286, 312, cf 336,
    cf 347

108 †Hemiauchenia macrocephala   36, 50, 52, cf 154, 190, ?
    194

109 Homo sapiens   26, 82, 117, 145, 154, 175, 191, 270,

298, 347

110 †Homotherium sp.   6

111 Lagurus curtatus   4, 9, cf 18, 26, 31, 91, 137, 143,
    145, 153, 154, cf 155, 162, 175, 176, 218, 242, 293,
    312, 334, 335, 347

112 †Lamini   52, 213

113 Lasionycteris noctivagans   26, 175, 324

114 Lasiurus cinereus   137, 248, 261

115 Leptonycteris nivalis   261

116 Leporidae   136, 281, 302

117 Lepus sp.   10, 22, 26, 31, 35, 46, 52, 79, 82, 86, 91,
    107, 118, 123, 138, 139, 149, 162, 202, 212, 218, 246,
    262, 270, 286, 293, 312, 318, 320, 335

118 Lepus americanus   155, 240, 259

119 Lepus californicus   3, cf 58, cf 85, 154, ? 157, 181,
    190, 228, ? 236, 240, cf 247, 248, 255, 257, cf 324,
    327, 345, 347

120 Lepus townsendi   cf 6, 9, 50, 135, 137, 145, 153, 154,
    cf 155, cf 175, cf 194, cf 196, cf 284, 347

121 Lepus townsendi or L. californicus   4, 143

122 Liomys irroratus   261

123 Lutra sp.   6

124 Lutra canadensis   294

125 Lynx sp.   58, 247

126 Lynx canadensis   155, 196, cf 284

127 Lynx rufus   3, 6, 9, 26, 46, 72, cf 82, 149, 153, 154, 155, 157, cf 162, cf 163, 175, 181, 202, 216, 240, 248, 255, 261, 279, 286, cf 345

128 Macrotus californicus   75

129 †Mammutidae   90, cf 223, 302

130 †Mammut sp.   138, 217

131 †Mammut americanum   69, 163, 169, 180, 190, 204, 240, 248, 260, 266, 308, 326

132 †Mammut or †Mammuthus   185, 191, 222

133 †Mammuthus sp.   4, 15, 21, 23, 24, 28, 29, 32, 36, 38, 39, 47, 48, 51, 52, 53, 60, 66, 67, 68, 70, 74, 96, 99, 102, 109, 110, 115, 119, 122, 124, 127, 128, 129, 133, 139, 145, 146, 150, 151, 152, 153, 159, 160, 165, 166, 167, 173, 174, 179, 183, 184, 189, 193, 194, 207, 211, 212, 213, 215, 218, 224, 232, 235, 237, 245, 246, 249,

251, 259, 263, 266, 269, 272, 274, 275, 280, 281, 303,
305, 306, 313, 329, 330, 337, 343

134 †Mammuthus columbi   6, cf 8, cf 19, 35, 44, 56, 57, 69,
79, 83, 85, 94, 108, 113, 120, ? 123, 141, ? 164, 169,
170, cf 180, 190, 196, 205, 214, 217, 248, cf 258, cf
260, 268, 273, cf 284, 292, 302, 318, 319, 320, 323,
poss 326, 342

135 †Mammuthus imperator   cf 6, 30, 112, 192, cf 199, 241,
326

136 †Mammuthus primigenius   57, 112, 178, 191, 195, 239,
240, 300

137 Marmota sp.   141, 202, 218, 261, 302

138 Marmota flaviventris   9, 22, 26, 46, 64, 72, 86, 89,
91, 107, 117, 131, 137, 139, 143, 154, cf 155, cf 162,
175, 182, 198, 200, 216, 221, 228, 240, 243, 247, 286,
304, 310, 311, 317, 324, cf 328, 347

139 Marmota or Erethizon   255

140 Martes americana   26, 72, 93

141 †Martes nobilis   26, 72, 83, 155, 175, 176, 218, 240,
259, 286, 347

142 †Megalonyx sp.   57, ? 138, 163, 180, ? 190, 196, 259,
260, 318, 320

143 †Megalonyx jeffersoni   6, 197, cf 222, ? 240, 246, 248, 262

144 Mephitinae (small skunk)   35

145 Mephitis sp.   cf 149

146 Mephitis macroura   261

147 Mephitis mephitis   4, 8, 9, 22, 26, 58, 72, 82, 138, 153, cf 154, 155, 175, 190, 228, 240, 248, cf 252, 259, 279

148 Microdipodops megacephalus   cf 286

149 Microtus sp.   22, 36, 74, 86, 107, 125, 126, 138, 147, 187, 194, 196, 200, 202, 210, 212, 218, 227, 263, 273, 278, 299, 318, 328, 335

150 Microtus californicus   58, 85, 162, cf 163, 190, 223, 240, 248, 259, cf 262, cf 320

151 Microtus longicaudus   4, 9, 31, cf 46, 91, 137, 175, 275, cf 286, cf 293, 347

152 Microtus longicaudus or M. montanus   72

153 Microtus longicaudus or M. pennsylvanicus   cf 6

154 Microtus mexicanus   9, 10, 31, cf 35, 46, 50, 82, 91, 137, 145, cf 149, 182, 216, ? 228, 236, 261, 293, 312, 324

155 Microtus montanus   cf 6, 10, 26, 123, 139, 145, cf
     155, 175, cf 219, cf 246, 277, cf 279, 284, cf 286, 347

156 Microtus montanus or M. californicus   127

157 Microtus pennsylvanicus   4, 9, 10, 26, 35, cf 108, cf
     136, 145, cf 153, 157, 175, 176, 216, 242, 275, 312,
     334

158 Mustela sp.   82, 135, 162, 202, 218, 259

159 Mustela erminea   72, cf 284, 286, ? 335, 347

160 Mustela frenata   9, 26, cf 50, 58, 72, 85, 137, 143,
     149, 155, 175, cf 180, 181, 182, 190, 216, 240, 248,
     259, 312, 324, cf 335, 347

161 Mustela nigripes   46, 72, 154, 155, 157, 175, 176, cf
     196, 225

162 †Mustela reliquus   261

163 Mustela vison   26, 72, 143, cf 154, 284, 286

164 †Mylodontidae   55, 281

165 Myotis sp.   9, 107, 162, 294, 312, 324

166 Myotis californicus   ? 137

167 Myotis evotis   175, ? 228

181 Neotoma fuscipes   138, 162, 190, 223, cf 262

182 Neotoma goldmani   ? 9, ? 10

183 Neotoma lepida   14, cf 46, 86, 127, 157, cf 162, 180,
     190, 247, cf 278, 286, 328

184 Neotoma mexicana   22, cf 46, 182, 216, cf 247, 324

185 Neotoma mexicana or N. cinerea   107

186 Neotoma micropus   ? 31, cf 137, ? 153, ? 154, 157,
     181, 182, 216, 312, 324

187 Neotoma stephensi   247, cf 328

188 †Nothrotheriops sp.   78, 82, 185, 186, 224, 266, 279,
     345, ? 347

189 †Nothrotheriops shastensis   3, 9, 52, 107, 135, 138,
     182, ? 193, 208, 222, 240, 244, 247, 248, 259, 261, 262,
     318, 324, 327

190 Notiosorex sp.   283, 350

191 Notiosorex crawfordi   10, 22, 31, 85, 107, 141, 145,
     157, 162, 182, 216, 223, 248, 279, 312, 324, 328

192 Ochotona sp.   296, 297

193 Ochotona princeps   26, 72, 86, cf 93, cf 125, cf 126,

132, 143, 155, 162, 175, 176, 200, 218, 242, cf 276, 286, cf 287, cf 288, 347

194 <u>Odocoileus</u> sp.   9, 22, 58, 72, 86, 108, 127, 138, 155, 160, 162, 182, 190, 193, 194, 196, 240, 248, 259, cf 265, 273, 279, ? 286, 293, 304, 320, 328

195 <u>Odocoileus</u> <u>hemionus</u>   3, cf 6, 26, 35, 46, 57, 118, cf 143, cf 163, 175, cf 180, cf 262, 345

196 <u>Odocoileus</u> <u>virginianus</u>   46, 345

197 <u>Odocoileus</u> or <u>Ovis</u>   126

198 <u>Ondatra</u> sp.   6, 85, 99, 284

199 <u>Ondatra</u> <u>zibethicus</u>   22, 26, 35, 36, 123, 127, 137, 143, 175, 196, 216, 246, 320

200 <u>Onychomys</u> sp.   40, ? 58, ? 190, 210, 348, 350

201 <u>Onychomys</u> <u>leucogaster</u>   9, 108, 137, 153, 154, 181, 182, 216, ? 228, 236, 312, cf 335

202 <u>Onychomys</u> <u>torridus</u>   153, 154, 182, 216, 248

203 <u>Oreamnos</u> sp.   57, 259, 310

204 <u>Oreamnos</u> <u>americanus</u>   26, ? 86, cf 143, 175, 176, 240

205 †<u>Oreamnos</u> <u>harringtoni</u>   247, 261, 286, 294, 309

206 †Oreamnos-like bovid   91, 216

207 Oreortyx pictus   46, 138, cf 145, cf 157, 240, 259,
     279

208 †Orthogeomys (Heterogeomys) onerosus   261

209 Ovibos sp. 80, 238

210 Ovibos moschatus   101, 109, 167, cf 336

211 Ovobovine indet.   26, 175, 242

212 Ovis sp.   75, 185, 270

213 Ovis aries   26, ? 86, 153, 154

214 Ovis canadensis   5, 9, 26, 37, 46, 52, 62, 72, 77, 86,
     87, 116, 127, 135, 143, 155, 162, 175, cf 176, 182, 196,
     202, 216, 218, 227, 231, cf 242, 247, 279, 286, 294,
     328, 345, 346

215 †Paleolama mirifica   180

216 †Panthera leo atrox   6, 16, 30, cf 58, 72, 153, 155,
     164, 167, 169, 175, 177, cf 180, 190, 196, 206, 216,
     218, ? 240, 248, 254, 261, cf 262, cf 273, 318, 327

217 Panthera onca   123, cf 236, 255, 261, 286, 290

218 Pappogeomys sp. 261

219 <u>Pappogeomys castanops</u>   9, 46, 91, 157, cf 181, ? 255, 261, 293, 324, 345

220 <u>Perognathus</u> sp.   9 (small), 31 (small), 58, 86, 91, 145 (small), 145 (large), 162, 175, 181, 212, cf 222, 270, 281, 312, 338, 348, 349, 350

221 <u>Perognathus californicus</u>   cf 85, cf 223, 248

222 <u>Perognathus flavus</u>   cf 153, 154, 216

223 <u>Perognathus flavescens</u>   ? 228

224 <u>Perognathus hispidus</u>   199

225 <u>Perognathus inornatus</u>   cf 190

226 <u>Perognathus intermedius</u>   154, cf 328, cf 345

227 <u>Perognathus longimembris</u>   127

228 <u>Perognathus parvus</u>   cf 286, cf 335

229 <u>Perognathus (Chaetodipus)</u>   255

230 <u>Perognathus (Perognathus)</u>   255

231 <u>Peromyscus</u> sp.   10, 14, 50, 72, 82, 86, 91, 107, cf 126, 143, 149, 153, 154, 155, 157, 162, 163, 175, 181, 182, 202, 210, 216, 218, 221, 247, 255, 281, 286, 294, 312, 314, 315, 324, 328, 335, 348, 349, 350

232 †Peromyscus anyapahensis   69

233 Peromyscus boylii   ? 31, 138, 261

234 Peromyscus boylii or P. truei   ? 228

235 Peromyscus californicus   cf 190, 253

236 †Peromyscus cragini   cf 199

237 Peromyscus crinitus   cf 31, ? 137, cf 223

238 Peromyscus difficilis   cf 137

239 Peromyscus eremicus   324

240 Peromyscus leucopus   31 (poss Holocene), cf 35

241 Peromyscus maniculatus   4, ? 9, 26, cf 31, cf 46, 85,
     cf 137, 139, 223, 228, 248, cf 255, 259, 284, 347

242 †Peromyscus nesodytes   69

243 †Peromyscus progressus   199

244 Peromyscus truei   cf 35, ? 347

245 Peromyscus (Haplomylomys)   58

246 Phenacomys sp.   335

247 Phenacomys intermedius   4, 26, 143, cf 175, cf 176, cf

242, 284, cf 286, 347

248 <u>Pipistrellus</u> sp.   162

249 <u>Pitymys</u> sp.   9, 91

250 <u>Pitymys</u> <u>ochrogaster</u>   26, 31, 35, 108, 175, 181, 199, 216, cf 236, 293, 312

251 †<u>Platygonus</u> sp.   36, 123, 180, ? 240, 248

252 †<u>Platygonus</u> <u>alemani</u>   228

253 †<u>Platygonus</u> <u>compressus</u>   4, 95, 108, cf 190, 273

254 †<u>Platygonus</u> <u>vetus</u>   cf 123

255 <u>Plecotus</u> sp.   31, 107, 181, 255

256 †<u>Plecotus</u> <u>tetralophodon</u>   261

257 <u>Plecotus</u> <u>townsendi</u>   cf 137, 182, 216, ? 228, 324

258 <u>Procyon</u> sp.   ? 270

259 <u>Procyon</u> <u>lotor</u>   35, 63, 138, 196, 248, 259

260 <u>Rangifer</u> sp.   57, 112, 196 (small)

261 <u>Rangifer</u> <u>tarandus</u>   77, 155, 196, 282

262 <u>Reithrodontomys</u> sp.   9, 31, 91, 145, 154, ? 190, 210

263 Reithrodontomys fulvescens   157, 216

264 Reithrodontomys humulis   cf 85

265 Reithrodontomys megalotis   cf 35, 162, 223, 248, 261

266 Reithrodontomys (Reithrodontomys)   255

267 Scapanus latimanus   138, 223, ? 240, 248

268 Sciuridae   281

269 Sciurus sp.   58, 182

270 Sciurus alleni   ? 261

271 Sciurus arizonensis   cf 35

272 Sciurus griseus   259

273 Sciurus or Tamiasciurus (tree squirrel)   218

274 Sigmodon sp.   145, 149, 157, 181, 210, 312

275 Sigmodon hispidus   35, 261

276 †Smilodon sp.   ? 138, 187, 246

277 †Smilodon fatalis   6, 36, ? 52, cf 85, cf 180, 193,
     194, 248, 261, cf 262, cf 284

278 Sorex sp.   72, 126, 175, cf 181, 246

279   Sorex cinereus   26, 35, 107, 182, 216, 261, 324

280   Sorex hoyi   175

281   Sorex merriami   9, 10, 31, 137, 216, 293, 312

282   Sorex monticolus/vagrans   9, cf 10, cf 22, 137, 141, 182, 216, 312

283   Sorex nanus   cf 9, 141

284   Sorex ornatus   cf 58, cf 190, 223, cf 248

285   Sorex palustris   26, 175, 216, 284

286   Sorex saussurei   261

287   Sorex trowbridgi   cf 58

288   Soricidae   286, 348

289   Spermophilus sp.   6, 82, 91, 107, 118, 126, 143, 145, 181, 202, 210, 218, 247, 260, ? 270, 302, 314, 337, 338

290   Speromphilus armatus   cf 284, 347

291   Spermophilus beecheyi   85, 138, 180, cf 222, 240, 248, 259, 262

292   Spermophilus beldingi   cf 286, ? 347

293   Spermophilus columbianus   335

294   Spermophilus lateralis   26, 72, 143, cf 155, cf 162,
      175, 200, 240, 259, cf 286, 327

295   Spermophilus richardsoni   26, 72, 108, 136, ? 137,
      139, 155, 194, 196, 246, 273, cf 286, 347

296   Spermophilus spilosoma   153, cf 157, 182, cf 261

297   Spermophilus tereticaudus   cf 162

298   Spermophilus townsendi   155, 335, 347

299   Spermophilus tridecemlineatus   4, 9, 26, 31, 137, 175,
      cf 199

300   Spermophilus tridecemlineatus or S. spilosoma   108,
      273

301   Spermophilus variegatus   46, 86, 107, 157, cf 162,
      175, 182, cf 190, 216, 228, 255, 324, 328, 345

302   Spermophilus (Itidomys)   255

303   Spilogale sp.   91, 202, cf 223

304   Spilogale putorius   9, 58, 82, 145, 155, 162, 175,
      181, 182, 190, cf 194, 216, 228, 236, 240, 248, 261,
      279, 286, 347

305 †Stockoceros sp.   82

306 †Stockoceros conklingi   ? 91, 216, 261, 279, 327

307 †Stockoceros onusrosagris   46, 216, 228

308 Sylvilagus sp.   10, 14, 22, 26, 72, 86, 107, 108, 143,
    145, 163, 175, 181, 182, 199, 202, 210, 218, 227, 244,
    246, 261, 270, 278, 286, 294, 314, 315, ? 318, 320, 324,
    328, 348, 350

309 Sylvilagus auduboni   ? 9, ? 46, cf 82, cf 85, 145,
    153, 154, 157, cf 162, 190, cf 222, 228, 236, 240, 248,
    255, 257, 259, 345

310 Sylvilagus bachmani   58, cf 85, 190, 223, 248, cf 262

311 Sylvilagus floridanus   cf 46, 153, 154, 196, cf 216

312 †Sylvilagus leonensis   261

313 Sylvilagus nuttalli   9, 31, 50, 91, 137, 145, 149,
    154, cf 155, 181, cf 216, 236, 255, 293, 312, 347

314 Sylvilagus nuttalli or S. floridanus   46, 82, cf 335

315 Sylvilagus nuttalli or S. auduboni   cf 4, cf 275, cf
    335

316 †Symbos sp.   ? 6, 92, 175, 176, 218, 242

317 †Symbos cavifrons   5, 33, 109, 120, 326

318 Synaptomys cooperi   261

319 Tadarida sp.   157, 216

320 Tadarida brasiliensis   181, 182, ? 228

321 †Tadarida constantinei   220

322 Tadarida macrotis   cf 181, 257

323 Tamias sp.   10, 58, 82, 86, 143, 155, 162, ? 181, 182, 202, 216, 218, 240, 314, 315

324 Tamias cinereicollis   ? 22

325 Tamias dorsalis   ? 228

326 Tamias minimus   cf 175, 284, 286, 335

327 Tamias umbrinus   cf 286

328 Tamiasciurus douglasi   240, 259

329 Tamiasciurus hudsonicus   107, 182, 216

330 Tamiasciurus (squirrel size of)   221

331 †Tapirus sp.   169, 180, 181, 213, 222, 248, 260, 327

332 †Tapirus californicus   352

333 Taxidea sp.   ? 6, 58, 82, 123, 240, cf 255

334 Taxidea taxus   9, 26, 46, 108, 120, cf 143, 154, 155, 162, 175, 187, 190, 202, 228, 248, 261, 268, cf 270, 279, 284, 286, 327

335   Tayassu sp.   cf 327

336   Tayassu tajacu   cf 91

337 †Tetrameryx sp.   65, 98, ? 320

338   Thomomys sp.   72, 86, 118, 126, 127, 138, 162, 175,
      202, 210, 221, 227, ? 270, 281, 314, 315, 348, 350

339   Thomomys bottae   9, 22, 31, 46, cf 58, 82, 85, 91,
      137, 145, 149, 153, 163, cf 180, cf 182, 190, cf 216, cf
      222, 223, 240, 248, 259, 262, 312, ? 320, 324

340   Thomomys bottae or T. umbrinus   ? 228

341 †Thomomys microdon   240, 259

342 †Thomomys scudderi   123

343   Thomomys talpoides   4, 9, 26, 31, 50, 91, 108, 137,
      149, cf 155, cf 194, cf 196, 273, 284, cf 286, 335, 339,
      347

344   Thomomys townsendi   cf 6, 246, 335

345   Thomomys umbrinus   46, cf 157, 261

346 †Thomomys vetus   123

347 †Tremarctos floridanus   261

348   Urocyon or Vulpes   145, 279

349 Urocyon sp.   82, 108, 270, 281

350 Urocyon cinereoargenteus   10, 50 cf 58, 154, 162, cf
     163, 181, 202, 228, 240, 248, 255, 257, 259, 345

351 Ursus sp.   cf 52, 138, cf 255, ? 286

352 Ursus americanus   3, 35, 145, 149, 153, 154, 157, 163,
     190, 228, 240, 248, 259, 261

353 Ursus americanus or †Arctodus   82

354 Ursus arctos   72, 155, 175, 248, cf 345

355 Vulpes sp.   123, 202, 236, 259, 293

356 Vulpes macrotis   153, 154, cf 155, cf 162, 190, 327

357 Vulpes velox   9, 26, 31, 36, 46, cf 50, 72, 82, 137,
     cf 149, 153, 154, cf 227, cf 257, 286, 312

358 Vulpes vulpes   4, 26, cf 35, 46, 60, 72, 143, 153,
     154, 155, 175, 196, 218, 240, 286

359 †Equus, Species A   50

360 †Leporidae, extinct rabbit   10, 157, 181, 255, 257

361 †Neotoma, Species A   181, 255

362 †Neotoma, Species B   181, 255

APPENDIX 3. DATA BASE SITES

Included here is a large number of sites, many not mentioned
elsewhere. The list is not complete, though most additional
sites are isolated finds or so general as to be of little
use. The main reasons for including sites not directly utilized in this study are two: 1) these data have influenced
my approach and concept of Late Pleistocene events, and 2)
the reader should have access to the same basic set of data
as I did to allow criticism and access to the sport of
second-guessing the author.

All the sites are believed to be Rancholabrean and it is
hoped that most or all are Sangamonian or later. Where more
specific age-related data are available, some comment has
been made. For some sites, elevation above sea level and
elevation ee have been given. No claim of consistency is
made here--where the information was readily available or
where I had some purpose in mind at the time, the calculations were made; otherwise they were not.

The general format is a site name, often in an abbreviated
form, followed by elevation in meters above sea level, elevation ee, any age comment, and by one or more citations (or
UTEP when there is material associated with the Vertebrate
Paleobiology Collections of the Laboratory for Environmental
Biology, University of Texas at El Paso). The citations are
meant to give entry into the literature, and are not exhaus-
tive. Sites that are treated directly in the text have the

taxa listed by name (taxa extinct in the study region marked by †); other sites have the taxa coded by number (from Appendix 2) to save space.

1 **Abiquiu,** Rio Arriba Co., NM. 1830 m, 5682 m ee. Simpson, 1963. **91**

2 **Acequia,** Minidoka Co., ID. McDonald, 1981. **19, 23**

3 **Aden Fumerole,** Dona Ana Co., NM. Ca. 1350 m, 4774 m ee 11,080 ± 200 (Y-1163B) on dung; (9,840 ± 160, Y-1163A, believed to be affected by preservatives) (Long and Martin, 1974). Non-sloth material may be Holocene. Lull, 1929; Simons and Alexander, 1964; UTEP). **34, 119, 127, 189, 195, 353**

4 **Agate Basin** Site, Niobrara Co., WY. 1190 m, 5845 m ee. Includes material from Clovis age (C) dated at 11,450 ± 110 (SI-3734), Folsom-age material (F) dated at 10,789 ± 120 (SI-3733) and 10,665 ± 85 (SI-3732), and Agate Basin cultural-level material (A) dated at 10,430 ± 570 (RL-557). The chronologically-later Hell Gap cultural level also is reported, but not recorded here. Walker, 1982b.
Antilocapra americana (F,A), † Bison cf antiquus (C,F,A), †Camelops sp. (C,F), Canis latrans (F), C. lupus (F,A), Canis sp. (F), Centrocercus urophasianus (F), Cervus elaphus (F), Clethrionomys gapperi (F), Lagurus curtatus (F,A), Lepus cf californicus or townsendi (F), †Mammuthus (C), Mephitis mephitis (F), Microtus longicaudus (C,F,A), M. pennsylvanicus (C,F), Peromyscus maniculatus (C,F), Phenacomys intermedius (C,F), †Platygonus compressus (F), Spermophilus tridecemlineatus (F), Sylvilagus cf nuttalli or auduboni (C,F), Thomomys talpoides (F,A), Vulpes vulpes (F)

5 **Alpine Fm**, Salt Lake Co., UT. 1330 m, 5717 m ee. Early to mid-Wisconsinan according to the literature, but probably actually late Wisconsinan (V. Geist, pers. comm.). Harington, 1968; Stock and Stokes, 1969. 214, 317

6 **American Falls**, Power, Bingham, and Bannock counties, ID. 1340 m, 5941 m ee. Probably Sangamonian. >32,000 (W-358) on charcoal beneath bison skull. Hopkins et al., 1969; Kurten and Anderson, 1980.

? Antilocapra americana, †Arctodus simus, †Bison alaskensis, †B. latifrons, †B. priscus, †Camelops cf hesternus, †C. cf huerfanensis, †C. cf minidokae, †Canis ? dirus, C. latrans, Castor sp., †Equus cf excelsus, †E. cf laurentius, †E. cf scotti, †Glossotherium harlani, †Homotherium sp., Lepus cf townsendi, Lutra sp., Lynx rufus, †Mammuthus cf columbi, †M. cf imperator, †Megalonyx jeffersoni, Microtus cf longicaudus or M. cf pennsylvanicus, M. cf montanus, Odocoileus cf hemionus, Ondatra zibethicus, †Panthera leo atrox, †Smilodon fatalis, Spermophilus sp., ? †Symbos sp., ? Taxidea sp., Thomomys cf townsendi

7 **Anaconda Pit**, Pima Co., AZ. Lindsay and Tessman, 1974. 72

8 **Anderson Basin/Circus Basin/Elephant Tusk Basin**, Roosevelt Co., NM. 1214 m. Late Wisconsinan. Stock and Bode, 1936; Wendorf and Hester, 1975. 17, 26, cf 77, cf 134, 147

9 **Animal Fair** (includes Charlies Parlor and Hampton Court), UTEP Loc. 22, Dry Cave, Eddy Co., NM. 1280 m, 4758 m ee. 15,030 ± 210 (I-6201) on collagen. Harris, 1977b; Harris and Porter, 1980; UTEP.

Bassariscus sp., Canis latrans, C. lupus, †Capromeryx sp., Cryptotis parva, Dipodomys spectabilis, †Equus conversidens, †E. niobrarensis, †E. cf occidentalis, Erethizon dorsatum, Felis concolor, †Hemiauchenia sp., Lagurus curtatus, Lepus townsendi, Lynx rufus, Marmota flaviventris, Mephitis mephitis, Microtus longicaudus, M. mexicanus, M. pennsylvanicus, Mustela frenata, Myotis spp., Neotoma albigula, N. cinerea, N. ? goldmani, †Nothrotheriops shastensis, Odocoileus sp., Onychomys leucogaster, Ovis candensis, Pappogeomys castanops, Perognathus sp. (small), Peromyscus ? maniculatus, Pitymys sp., Reithrodontomys sp., Sorex merriami, S. monticolus/vagrans, S. cf nanus, Spermophilus tridecemlineatus, Spilogale putorius, Sylvilagus auduboni/floridanus, S. nuttalli, Taxidea taxus, Thomomys bottae, T. talpoides, Vulpes velox

10 **Anthony Cave**, UTEP Loc., 29, Dona Ana Co., NM, and El Paso Co., TX. 1555 m, 4979 m ee. Smartt, 1977; UTEP. 15, cf 34, 54, 72, 117, 154, 155, 157, 176, 179, ? 182, 191, 231, 281, cf 282, 308, 323, 350, 360

11 **Arikaree River**, CO. McDonald, 1981. 21

12 **Arizpe**, Sonora, Mexico. Rea, 1980. cf 20, cf 85

13 **Arroyo Las Positas**, Alameda Co., CA. McDonald, 1981; Nowak, 1979. 21, 31

14 **Artillery Mts.**, Mohave Co., AZ. Midden A. 615 m. 10,250 ± 200 (A-1099). Midden C. 725 m. >30,000 (A-1100). Van Devender and King, 1971. 13, 183, 231, 308

15 Ash Canyon, Cochise Co., AZ. 1890 m. Lindsay and
Tessman, 1974. 133

16 **Astor Pass,** Washoe Co., NV. 16,800 ± 600 on marl
(L-364cr); 17,500 ± 600 on shell from L-364cr (L-364cs)
(Hester, 1960). Merriam, 1915. cf 26, 72, 216

17 **Athabasca,** Alberta. McDonald, 1981. 21

18 **Atlatl Cave,** San Juan Co., NM. 1910 m, 5762 m ee. Bulk
of fauna ca. 4855 ± 70 (DIC-591) (Gillespie, pers. comm.),
but Betancourt and Van Devender (1981) date woodrat middens
in the shelter as 5550 ± 130 (A-2115); 9460 ± 160 (A-2116);
10,030 ± 150 (A-2123); 10,500 ± 250 (A-2411); 10,600 ± 200
(A-2139). Lagurus suggested as earlier than rest of fauna.
Gillespie, pers. comm., 1981.
Lagurus cf curtatus

19 **Babine Lake,** B.C. 42,900 ± 1860 (GSC-1657) on wood;
43,800 ± 1830 (GSC-1687) on wood; 34,000 ± 690 (GSC-1754) on
mammoth bone. Harington, 1978. cf 134

20 **Badger Springs,** AZ. McDonald, 1981. 21

21 **Bain** Site, Pinal Co., AZ. Lindsay and Tessman, 1974.
133

22 **Baldy Peak Cave,** UTEP Loc. 94, Luna Co., NM. 1550 m, 4974
m ee. Late Wisconsinan. UTEP. 14, 117, 138, 147, 149, 179,
184, 191, 194, 199, cf 282, 308, ? 324, 339

23 **Bear Creek**, Goshen Co., WY. Walker and Frison, 1980. 133

24 **Bear Lake Region**, Rich Co., UT. Miller, 1976. 133

25 **Beaumont**, Riverside Co., CA. McDonald, 1981. 21

26 **Bell Cave**, Albany Co., WY. 2379 m, 6821 m ee. Late Wisconsinan. Anderson, 1974; Walker, 1982b; Zeimens and Walker, 1974.
Antilocapra americana, Bison bison, †Camelops cf hesternus, Canis latrans, C. lupus, Cervus elaphus, Cynomys cf gunnisoni, Dicrostonyx torquatus, Eptesicus fuscus, Equus ? caballus, †E. conversidens, †Equus sp. (large), Erethizon dorsatum, Gulo gulo, Homo sapiens, Lagurus curtatus, Lasionycteris noctivagans, Lepus sp., Lynx rufus, Marmota flaviventris, Martes americana, †Martes nobilis, Mephitis mephitis, Microtus montanus, M. pennsylvanicus, Mustela frenata, M. vison, Myotis lucifugus, Neotoma cinerea, Ochotona princeps, Odocoileus hemionus, Ondatra zibethicus, Oreamnos americanus, Ovis canadensis, O. cf aries, Ovobovini indet., Peromyscus maniculatus, Phenacomys intermedius, Pitymys ochrogaster, Sorex cinereus, S. palustris, Spermophilus lateralis, S. richardsoni, S. tridecemlineatus, Sylvilagus sp., Taxidea taxus, Thomomys talpoides, Vulpes velox, V. vulpes

27 **Bend**, Deschutes Co., OR. Late Pleistocene. Nowak, 1979. 35

28 **Bentzen-Kaufmann Cave**, Sheridan Co., WY. >6975 ± 275 (Don Gray date). Anderson, 1974. 133

29 Big Tooth, Cochise Co., AZ. Lindsay and Tessman, 1974.
133

30 Bindloss Area, Alberta. 650 m. Post-Wisconsinan age of
>8000 suggested by Churcher (1972), but Harington (1978)
suggests probably interglacial or interstadial. Churcher,
1972. 76, 135, 216

31 Bison Chamber, UTEP Loc. 4, Dry Cave, Eddy Co., NM. 1280
m, 4758 m ee. <14,470, >10,730, interpolated from ^{14}C dates.
Harris, 1977b, UTEP.
Antilocapra americana, †Bison cf antiquus, Dipodomys cf spectabilis, Eptesicus fuscus, †Equus cf conversidens, †E. niobrarensis, Lagurus curtatus, Lepus sp., Microtus longicaudus,
M. mexicanus, †Myotis cf rectidentis, M. velifer, Neotoma
cinerea, N. ? micropus, Notiosorex crawfordi, Perognathus sp.
(small), Peromyscus ? boylii, P. cf crinitus, P. leucopus, P.
cf maniculatus, Pitymys ochrogaster, Plecotus sp., Reithrodontomys sp., Sorex merriami, Spermophilus tridecemlineatus,
Sylvilagus nuttalli, Thomomys bottae, T. talpoides, Vulpes
velox

32 Black Mountain, Yavapai Co., AZ. Lindsay and Tessman,
1974. 133

33 Black Rocks, McKinley Co., NM. Ca. 2020 m, 5765 m ee.
Allen, 1913. 317

34 Blackwater Draw, Roosevelt Co., NM. Hester, 1967.
†Capromeryx minor

35 Blackwater Loc. No. 1, Brown Sand Wedge, Roosevelt Co.,

NM. Ca. 1280 m, 4972 m ee. 11,170 ± 360 (A-481) (Haynes, 1964). Slaughter, 1975.

†Bison cf antiquus, Canis lupus, †Capromeryx sp., Cynomys ludovicianus, †Dasypus bellus, Didelphis virginianus, †Equus niobrarensis, Geomys cf bursarius, Lepus sp., †Mammuthus columbi, Mephitinae (small skunk), Microtus cf mexicanus, M. pennsylvanicus, Neotoma sp., Odocoileus hemionus, Ondatra zibethicus, Peromyscus cf leucopus, P. cf truei, Pitymys ochrogaster, Procyon lotor, Reithrodontomys cf megalotis, Sciurus cf arizonensis, Sigmodon hispidus, Sorex cinereus, Ursus americanus, Vulpes cf vulpes

36 **Blackwater Loc. No. 1, Gray Sand**, Roosevelt Co., NM. ca. 1280 m, 4972 m ee. Age probably >15,000 (Haynes and Agogino, 1966). Harris and Porter, 1980; Lundelius, 1972b.
Antilocapridae, †Bison antiquus, †Camelops sp., Canis latrans, C. lupus, †Equus conversidens, †E. niobrarensis, Geomys sp., †Hemiauchenia macrocephala, †Mammuthus sp., Microtus sp., Ondatra zibethicus, †Platygonus sp., †Smilodon fatalis, Vulpes velox.

37 **Bloomfield**, San Juan Co., NM. Ca. 1645 m, 5551 m ee. Stovall, 1946. **214**

38 **Boquillas Station**, Cochise Co., AZ. 1183 m. Lindsay and Tessman, 1974. ? **17, 72, 133**

39 **Bosler Gravel Pit**, Albany Co., WY. Anderson, 1974. **133**

40 **Brass Cap Pt.**, Yuma Co., AZ. 550 m. Lindsay and Tessman, 1974. **175, 200**

41 Brewster Site, Niobrara Co., WY.   1190 m, 5845 m ee.
Folsom Level, 10,375 ± 700 (I-472). Haynes, 1964. 21

42 Bridger, Uinta Co., WY. Anderson, 1974. 72

43 Brophy Cienega, nr. Elgin, Santa Cruz Co., AZ. McDonald,
1981. 21

44 Buffalo, Johnson Co., WY. Anderson, 1974. 134

45 Burke Ranch, Contra Costa Co., CA. McDonald, 1981. 21

46 Burnet Cave, Eddy Co., NM. 1403 m, 4881 m ee. 7432 ± 300
(solid carbon date, C-823) questioned (Hester, 1967); includes Holocene material. Schultz and Howard, 1935.
Antilocapra cf americana, †Arctodus sp., Bassariscus astutus,
†Bison antiquus, †Camelops sp., Canis latrans, C. lupus,
Conepatus cf mesoleucus, Cynomys ludovicianus, Dipodomys
ordi, †Equus conversidens, †E. niobrarensis, †E. cf tau,
†Euceratherium collinum, Felis concolor, Lepus sp., Lynx
rufus, Marmota flaviventris, Microtus cf longicaudus, M.
mexicanus, Mustela nigripes, †Navahoceros fricki, Neotoma cf
cinerea, N. cf lepida, N. cf mexicana, Odocoileus hemionus,
O. virginianus, Oreortyx pictus, Ovis canadensis, Pappogeomys
castanops, Peromyscus cf maniculatus, Spermophilus variegatus, †Stockoceros onusrosagris, Sylvilagus ? auduboni, S.
floridanus or S. nuttalli, Taxidea taxus, Thomomys bottae, T.
umbrinus, Vulpes velox, V. vulpes

47 Calera Ranch, Pima Co., AZ. Lindsay and Tessman, 1974.
133

48 Calgary, Alberta. Harington, 1978. 21, cf 27, 76, 133

49 California Sand and Gravel Co. Pit, Alameda Co., CA. McDonald, 1981. 19

50 Camel Room, UTEP Loc. 25, Dry Cave, Eddy Co., NM. 1280 m, 4758 m ee. Harris, 1977b; UTEP.
Bison sp., †Camelops ? hesternus, Conepatus mesoleucus, †Equus niobrarensis, †Equus Species A, †Hemiauchenia macrocephala, Lepus townsendi, Microtus mexicanus, Mustela cf frenata, Neotoma cinerea, Peromyscus sp., Sylvilagus nuttalli, Thomomys talpoides, Urocyon cinereoargenteus, Vulpes cf velox

51 Cameron, Coconino Co., AZ. Lindsay and Tessman, 1974. 133

52 Camp Cady L. F., Manix Lake, San Bernardino Co., CA. George T. Jefferson (pers. comm.). ? 7, 11, 21, cf 27, aff. ? 29, 31, ? 34, 73, ? 76, 96, 104, 108, 112, 117, 133, 189, 214, ? 277, cf 351

53 Canez Wash No. 1, Pima Co., AZ. Lindsay and Tessman, 1974. 72, 133

54 Canez Wash No. 2, Pima Co., AZ. Lindsay and Tessman, 1974. 72

55 Canyon De Chelly, Apache Co., AZ. Lindsay and Tessman, 1974. 164

56 Careyhurst, Converse Co., WY. Anderson, 1974. 134

57 **Cariboo District**, B.C. (includes Quesnel Forks). <u>Oreamnos</u>

cf. Sangamonian or earlier; dates not established for others.
Harington, 1978. **17, 87, 134, 136, 142, 195, 203, 260**

58 **Carpinteria**, Santa Barbara Co., CA. 4 m, 3696 m ee.
Wilson, 1933.
? <u>Bison</u> sp., ? †<u>Camelops</u> cf <u>hesternus</u>, †<u>Canis</u> cf <u>dirus</u>, <u>C.</u>
<u>latrans</u>, <u>Dipodomys</u> sp., †<u>Equus</u> cf <u>occidentalis</u>, <u>Geococcyx</u>
<u>californianus</u>, <u>Lepus</u> cf <u>californicus</u>, <u>Lynx</u> sp., <u>Mephitis</u>
<u>mephitis</u>, <u>Microtus</u> <u>californicus</u>, <u>Mustela</u> <u>frenata</u>, <u>Neotoma</u>
sp., <u>Odocoileus</u> sp., ? <u>Onychomys</u> sp., †<u>Panthera</u> cf <u>leo</u> <u>atrox</u>,
<u>Perognathus</u> sp., <u>Peromyscus</u> (<u>Haplomylomys</u>) sp., <u>Sciurus</u> sp.,
<u>Sorex</u> cf <u>ornatus</u>, <u>S.</u> cf <u>trowbridgi</u>, <u>Spilogale</u> <u>putorius</u>, <u>Syl-</u>
<u>vilagus</u> <u>bachmani</u>, <u>Tamias</u> sp., <u>Taxidea</u> sp., <u>Thomomys</u> cf <u>bot-</u>
<u>tae</u>, <u>Urocyon</u> cf <u>cinereoargenteus</u>

59 **Carter/Kerr-McGee** Site, Campbell Co., WY. Latest
Pleistocene. Frison et al., 1978. **17, 26**

60 **Casper**, Natrona Co., WY. ^{14}C date of 10,080 ± 170
(RL-208) on bone, 9830 ± 350 (RL-125) on charcoal. Frison et
al., 1978. **21, 26, 133, 358**

61 **Castleguard Icefield**, Banff Natl. Park, Alberta. Late
Pleistocene. Nowak, 1979. **31**

62 **Catclaw Cave**, Mohave Co., AZ. Hibbard and Wright, 1956.
214

63 **Cave Bear Cave**, Shasta Co. ?, CA. Arata and Hutchison,
1964. **259**

64 Cave in Manzano Mts., Torrance Co., NM. Howell, 1915.
138

65 Ceremonial Cave, Hueco Mts., El Paso Co., TX. Cosgrove,
1947. ? 78, 107, 337

66 Cerro Colorado No. 1, Pima Co., AZ. Lindsay and Tessman,
1974. 72, 133

67 Cerro Colorado No. 2, Pima Co., AZ. Lindsay and Tessman,
1974. 133

68 Cerros Negros, Pinal Co., AZ. 945 m. 12,000 ± 300
(A-854) (Haynes, 1968). Lindsay and Tessman, 1974. 17, 72,
133

69 Channel Islands (incl. Santa Rosa Is.), CA. 11,800 (UCLA-
106) on charcoal for Mammuthus (Hester, 1967); also 29,650 ±
2500 (L-290R) according to Orr (1956); 29,700 ± 3000 (L-290R)
according to Hester (1960) on charred mammoth bone, 36-ft
level; 16,700 ± 1500 on charcoal from middle of mammoth beds
(M-599); and 15,820 ± 280 (L-244) on wood for mammoth level
(Hester, 1960). Kurten and Anderson, 1980.
†Mammut americanum, †Mammuthus columbi, †Peromyscus anyapahensis, †P. nesodytes

70 Charley Day Spr., Coconino Co., AZ. Lindsay and Tessman,
1974. 21, 26, 72, 133

71 Cheyenne, Laramie Co., WY. Anderson, 1974. 72

72 Chimney Rock Animal Trap, Larimer Co., CO. Hager, 1972.

Antilocapra americana, Bison bison, Canis latrans, C. lupus, Cynomys sp., Erethizon dorsatum, Felis concolor, Gulo gulo, Lynx rufus, Marmota flaviventris, Martes americana, †M. nobilis, Mephitis mephitis, Microtus longicaudus or M. montanus, Mustela erminea, M. frenata, M. nigripes, M. vison, Neotoma cinerea, Ochotona princeps, Odocoileus sp., Ovis canadensis, †Panthera leo atrox, Peromyscus sp., Sorex sp., Spermophilus lateralis, S. richardsoni, Sylvilagus sp., Thomomys sp., Ursus arctos, Vulpes velox, V. vulpes

73 **China Lake,** Kern Co., CA.   18,600 ± 4500 on ivory.   Kurten and Anderson, 1980.   21

74 **Choate Ranch,** Cochise Co., AZ.   1234 m.   Lindsay and Tessman, 1974.   25, 72, 133, 149

75 **Cinnabar Mine,** Brewster Co., TX.   Cockerell, 1930; Ray and Wilson, 1979.   17, 60, 72, 128, 212

76 **Clayhurst Crossing,** B.C.   Churcher and Wilson, 1979.   17

77 **Cochrane** (Griffin South Pit, Griffin North Pit, Clark Pit--Bughill Creek Fm), Alberta.   Griffin Pit, 10,760 ± 160 (GSC-612); Clark Pit, 11,370 ± 170 (GSC-613); 11,100 ± 160 (GSC-989).   Wilson and Churcher, 1978.
†Bison antiquus, Cervus elaphus, †Equus conversidens, Ovis canadensis, Rangifer tarandus

78 **Coconino Cavern,** Coconino Co., AZ.   Lindsay and Tessman, 1974.   188

79 **Colby Site/Worland Area,** Washakie Co., WY.   11,200 ± 200

(RL-392) on mammoth bone. Frison et al., 1978; Mears, 1981; Walker and Frison, 1980. **8, 22, 26, 76, 117, 134**

80 **Cold Lake**, Alberta. Harington, 1978. **209**

81 **Comondu**, Baja California Sur, Mexico. McDonald, 1981. **21**

82 **Conkling Cavern**, Dona Ana Co., NM. 1399 m, 4823 m ee. Conkling, 1932; Nowak, 1979; Smartt, 1977; pers. obs. Antilocapra sp., Bassariscus sp., Bison sp., †Camelops sp., †Canis dirus, C. latrans, †Capromeryx sp., Centrocercus urophasianus, Cynomys cf ludovicianus, Dipodomys sp., †Equus sp., Felis concolor, †Geococcyx californianus conklingi, Geomys sp., †Hemiauchenia sp., Homo sapiens, Lepus sp., Lynx cf rufus, Mephitis mephitis, Microtus mexicanus, Mustela sp., Neotoma albigula, N. cinerea, †Nothrotheriops sp., Peromyscus sp., Spermophilus sp., Spilogale putorius, †Stockoceros sp., Sylvilagus cf auduboni, S. nuttalli or S. floridanus, Tamias cf cinereicollis, Taxidea sp., Thomomys bottae, Urocyon sp., Ursus americanus or †Arctodus sp., Vulpes velox

83 **Converse Co.**, WY. Anderson, 1974. **12, 17, 72, 134, 141**

84 **Cool Quarry**, Eldorado Co., CA. Late Pleistocene. McDonald, 1981; Nowak, 1979. **19, 31**

85 **Costeau Pit**, Los Angeles Co., CA. 90 m. >40,000 BP. Miller, 1971. **21, 23, cf 27, cf 31, cf 34, 36, 43, 63, cf 74, cf 82, cf 97, cf 105, 107, cf 119, 134, 150, 160, 175, 191, 198, cf 221, 241, cf 264, cf 277, 291, cf 309, cf 310, 339**

86 Council Hall Cave, Smith Creek Canyon, White Pine Co., NV.
2040 m, 6213 m ee. 23,900 ± 970 (GaK-970); 13,040 ± 190
(WK-161). Thompson and Mead, 1982. ? 7, cf 34, 117, 138,
149, 179, 183, 193, ? 194, ? 204, ? 213, 214, 220, 231, 301,
308, 323, 338

87 Crypt Cave, Pershing Co., NV. Ca. 1240 m, 5520 m ee.
19,750 ± 650 (L. 364-BS) (Orr, 1969). Orr, 1969, 1972. 1,
25, 72, 214

88 Cueva Las Cruces, Dona Ana Co., NM. Kurten, 1975. 174

89 Cylinder Cave, Coconino Co., AZ. Ca. 1400 m. Lange,
1956. 138

90 Dam (Hop-Strawn Pit), Power Co., ID. Kurten and Anderson,
1980; McDonald, 1981.
†Bison antiquus, †Camelidae, †Edentata (ground sloths),
†Equus sp., †Mammutidae, †Mammuthus sp.

91 Dark Canyon Cave, UTEP Loc. 75, Eddy Co., NM. 1100 m,
4578 m ee. Harris and Porter, 1980; Howard, 1971; Lundelius,
1979; Nowak, 1979; UTEP.
Antrozous pallidus, †Canis cf dirus, C. latrans, C. lupus,
†Capromeryx sp., Cynomys sp., Dipodomys cf spectabilis,
†Equus conversidens, †E. niobrarensis, †Geococcyx californianus conklingi, Lagurus curtatus, Lepus sp., Marmota flaviventris, Microtus longicaudus, M. mexicanus, Neotoma albigula, N. cinerea, †Oreamnos-like bovid, Pappogeomys castanops,
Perognathus sp., Peromyscus sp., Pitymys sp., Reithrodontomys
sp., Sorex sp., Spermophilus sp., Spilogale sp., ? †Stocko-

ceros conklingi, Sylvilagus nuttalli, cf Tayassu tajacu,
Thomomys bottae, T. talpoides

92 **Dease Lake**, B.C.   Harington, 1968.   316

93 **Deer Creek Cave**, Jarbridge Mts., Elko Co., NV.   1770 m,
6264 m ee.   10,085 ± 400 (I-1028); 9670 ± 300 (I-1029).
Thompson and Mead, 1982.   **140, cf 193**

94 **Dent**, Weld Co., CO.   11,200 ± 500 (I-622) on mammoth bone
(Haynes, 1964).   Kurten and Anderson, 1980.   **134**

95 **Denver**, Denver Co., CO.   Kurten and Anderson, 1980.   **21,
28, 253**

96 **Donnet**, Cochise Co., AZ.   Lindsay and Tessman, 1974.   **133**

97 **Doolan Canyon**, Alameda Co., CA.   McDonald, 1981.   **21**

98 **Doolittle Cave**, Grant Co., NM.   Cosgrove, 1947.   **36, 337**

99 **Double Adobe**, Cochise Co., AZ.   Lindsay and Tessman, 1974.
17, 26, 34, 133, 198

100 **Douglas**, Cochise Co., AZ.   Lindsay and Tessman, 1974.   **25**

101 **Douglas**, Converse Co., WY.   1468 m, 6069 m ee.   Walker,
1982a.
Ovibos moschatus

102 **Dover**, Albany Co., WY.   Anderson, 1974.   **133**

103 **Dragoon Mts.**, Cochise Co., AZ. Lindsay and Tessman, 1974. 72

104 **Drumheller**, Alberta. McDonald, 1981. 21

105 **Duck Point**, Power Co., ID. Kurten and Anderson, 1980. 106

106 **Dunniway Gravel Pit**, Malheur Co., OR. Late Pleistocene. Allison and Boyd, 1954. cf 26

107 **Dust Cave**, Culberson Co., TX. Ca. 2000 m, 5424 m ee. Logan, pers. comm.; Van Devender et al., 1977. Antrozous pallidus, Canis sp. (large), Cryptotis parva, Eptesicus fuscus, Erethizon dorsatum, Lepus sp., Marmota flaviventris, Microtus sp., Myotis sp., M. velifer, Neotoma albigula or N. micropus, N. mexicana or N. cinerea, †Nothrotheriops shastensis, Notiosorex crawfordi, Peromyscus spp., Plecotus sp., Sorex cinereus, Spermophilus sp., S. variegatus, Sylvilagus sp., Tamiasciurus hudsonicus

108 **Dutton**, Yuma Co., CO. 1255 m, 5482 m ee. Below Clovis cultural level, >11,800, <29,000. Graham, 1981. Antilocapra sp., †Bison antiquus, †Camelops hesternus, Cynomys ludovicianus, †Equus sp. (larger), †Equus sp. (smaller), Geomys bursarius, †Glossotherium harlani, †Mammuthus columbi, Microtus cf pennsylvanicus, Odocoileus sp., Onychomys leucogaster, Pitymys ochrogaster, †Platygonus compressus, Spermophilus richardsoni, S. spilosoma or S. tridecemlineatus, Sylvilagus sp., Taxidea taxus, Thomomys talpoides, Urocyon sp.

109 **Edmonton Area**, Alberta. Harington indicates several ages

mixed, including pre-glaciation. Harington, 1978; McDonald, 1981. **19, 21, 27, 76, 133, 210, 317**

110 **Elgin School**, Santa Cruz Co., AZ. Lindsay and Tessman, 1974. **133**

111 **Empire South**, Pima Co., AZ. Lindsay and Tessman, 1974. **72**

112 **Empress**, Alberta. Ca. 640 m, 6097 m ee. 14,200 ± 1120 (GSC-1199); 20,400 ± 320 (GSC-1387) on bone fragments. Harington, 1978.
†<u>Bison</u> cf <u>antiquus</u>, †<u>Camelops</u> cf <u>hesternus</u>, †<u>Equus</u> <u>conversidens</u>, †<u>E</u>. <u>scotti</u>, †<u>Mammuthus</u> <u>imperator</u>, †<u>M</u>. <u>primigenius</u>, <u>Rangifer</u> sp.

--- **Entrance Chamber**, UTEP Loc. 24, Dry Cave, Eddy Co., NM. 1280 m, 4758 m ee. Holocene. UTEP.
<u>Dipodomys</u> sp. (small), <u>Homo</u> <u>sapiens</u>, <u>Lepus</u> sp., <u>Neotoma</u> <u>albigula</u>, <u>N</u>. <u>mexicana</u>, <u>Microtus</u> cf <u>mexicanus</u>, <u>Notiosorex</u> <u>crawfordi</u>, <u>Onychomys</u> sp., <u>Perognathus</u> cf <u>hispidus</u>, <u>Perognathus</u> sp. (small), <u>Peromyscus</u> sp., <u>Sigmodon</u> <u>hispidus</u>, <u>Sylvilagus</u> sp., Vespertilionidae

113 **Escapule**, Cochise Co., AZ. 1273 m. Lindsay and Tessman, 1974. **25, 30, 72, 134**

114 **Etna Cave**, Lincoln Co., NV. Frison et al., 1978; Jelinek, 1957. **26, 69, 72**

115 **Fenn Site**, Cochise Co., AZ. Lindsay and Tessman, 1974. **26, 72, 133**

116 Finlay Forks, B.C.   9280 ± 200 (GSC-1497) on bone.
Harington, 1978.   214

117 Fishbone Cave, Pershing Co., NV.   1238 m, 5518 m ee.
11,400 ± 250 (L-245) lowest occupation level (questioned--see
Hester, 1960); date on vegetal material from cultural stratum
lying on lake bed, 11,555 ± 500; later run, 10,900 ± 300
(Orr, 1956).   Orr, 1956.   26, 72, 109, 138

118 Folsom Site, Union Co., NM.   Folsom age.   Hay and Cook,
1930.   21, 57, 117, 195, 289, 338

119 Fort McDowell, Maricopa Co., AZ.   Lindsay and Tessman,
1974.   133

120 Fort Qu'Appelle (Bliss Gravel Pit-Echo Lake Gravel),
Saskatchewan.   Sangamonian (cf. Wisconsinan interstadial).
Harington, 1978.
†Arctodus simus, †Bison latifrons, †Camelops cf hesternus,
Canis lupus, †Cervalces scotti, †Equus cf conversidens, †E.
scotti, †Mammuthus columbi, Taxidea taxus, †Symbos cavifrons

121 Fort Saskatchewan, Alberta.   Churcher, 1968.   76

122 Fort Steele Site, Carbon Co., WY.   "Likely Late
Pleistocene."   Anderson, 1974.   133

123 Fossil Lake, Lake Co., OR.   1310 m, 5911 m ee.   Probably
mid or possibly early Wisconsinan (Allison, 1966).   Elftman,
1931; Russell, 1968.
Antilocapra cf americana, †Arctodus simus, †Camelops hesternus, †Canis cf dirus, C. latrans, C. cf lupus, Canis sp.

(smaller than C. latrans), Castor cf canadensis, Centrocercus urophasianus, †Equus pacificus, †Equus sp. (small), Felis sp. (small), †Glossotherium cf harlani, ? †Hemiauchenia sp., Lepus sp., †Mammuthus ? columbi, Microtus montanus, Ondatra zibethicus, Panthera onca, †Platygonus cf vetus, †Platygonus sp. (small), Taxidea sp., †Thomomys scudderi, †T. vetus, Vulpes sp.

124 **Gardner Gravel**, Cochise Co., AZ. Lindsay and Tessman, 1974. 133

125 **Garrison No. 1**, Millard Co., UT. 1640 m, 5813 m ee.
12,230 ± 180 (A-2312). Mead et al., 1982.
Brachylagus idahoensis, Microtus sp., Neotoma sp., Ochotona princeps

126 **Garrison No. 2**, Millard Co., UT. 1640 m, 5813 m ee.
13,480 ± 250 (A-2313). Mead et al., 1982.
Brachylagus idahoensis, †Camelops cf hesternus, Microtus sp., Neotoma sp., Ochotona cf princeps, Ovis or Odocoileus, cf Peromyscus, Sorex sp., Spermophilus sp., Thomomys sp.

127 **Glendale**, Clark Co., NV. 555 m, 4461 m ee. Van Devender and Tessman, 1975. 25, 39, 72, 133, 156, 183, 194, 199, 214, 227, 338

128 **Glenrock**, Converse Co., WY. Walker and Frison, 1980.
133

129 **Goodwater Wash**, Navajo Co., AZ. Lindsay and Tessman, 1974. 133

130 **Gottville**, Siskyou Co., CA. McDonald, 1981. 21

131 **Government Cave**, Coconino Co., AZ. Ca. 2135 m. Lange,
1956. 138

132 **Granite Canyon No. 1**, Deep Creek Range, Juab Co., UT.
2070 m, 6350 m ee. 13,620 ± 700 (A-2432). Thompson and
Mead, 1982.
Ochotona cf princeps

133 **Gray Site**, Cochise Co., AZ. 1128 m. Lindsay and
Tessman, 1974. 133

134 **Greybull**, Big Horn Co., WY. Anderson, 1974. 27

135 **Gypsum Cave**, Clark Co., NV. 610 m, 4462 m ee. See comments on dating by Hester (1960). More recent dates in Long
and Martin (1974): 11,690 ± 250 (IJ-452) and 11,360 ± 260
(A-1202). Harrington, 1933.
†Camelops sp., Canis lupus or †C. dirus, †Equus cf conversidens, †E. cf occidentalis, †Hemiauchenia sp., Lepus townsendi, Mustela sp., †Nothrotheriops shastensis, Ovis canadensis

136 **Hand Hills**, Alberta. Mid to Late Pleistocene. Prob.
Irvingtonian or later. Harington, 1978. 57, 70, cf 76, 101,
116, cf 157, 295

137 **Harris' Pocket**, UTEP Loc. 6, Dry Cave, Eddy Co., NM.
1280 m, 4758 m ee. 14,470 ± 250 (I-3365). Harris, 1977b,
UTEP.
? Antilocapra sp., Bison sp., Canis latrans, Cynomys (Leuco-

crossuromys) sp., Eptesicus fuscus, †Equus conversidens, †E.
niobrarensis, Erethizon dorsatum, Lagurus curtatus, Lasiurus
cinereus, Lepus townsendi, Marmota flaviventris, Microtus
longicaudus, M. mexicanus, Mustela frenata, Myotis ? californicus, M. lucifugus, †M. ? rectidentis, M. velifer, Neotoma
albigula, N. cinerea, N. cf micropus, Ondatra zibethicus,
Onychomys leucogaster, Peromyscus ? crinitus, P. cf difficilis, P. cf maniculatus, Plecotus cf townsendi, Sorex merriami, S. monticolus/vagrans, Sylvilagus nuttalli, Spermophilus
? richardsoni, S. tridecemlineatus, Thomomys bottae, T. talpoides, Vulpes velox

138 **Hawver Cave**, Eldorado Co., CA.   393 m, 4566 m ee.   Stock,
1918.
Aplodontia rufa, Bison sp., †Canis cf dirus, C. latrans,
†Equus ? occidentalis, †Euceratherium collinum, Felis concolor, †Glossotherium harlani, Lepus sp., †Mammut sp., ?
†Megalonyx sp., Mephitis mephitis, Microtus sp., Neotoma fuscipes, †Nothrotheriops shastensis, Odocoileus sp., Oreortyx
pictus, Peromyscus boylii, Procyon lotor, Scapanus latimanus,
? †Smilodon sp., Spermophilus beecheyi, Thomomys sp., Ursus
sp.

139 **Hell Gap Site**, Sec. 10, T28N, R65W, in Goshen Co., WY.
1525 m, 6073 m ee. <11,000 BP (Anderson, 1974); ca. 13,000
BP (Mears, 1981).   Anderson, 1974; Mears, 1981; Roberts,
1970.
Canis sp., Erethizon dorsatum, Lepus sp., †Mammuthus sp.,
Marmota flaviventris, Microtus montanus, Peromyuscus maniculatus, Spermophilus richardsoni

140 **Hereford Dairy**, Cochise Co., AZ. 1283 m, 4654 m ee.

Lindsay and Tessman, 1974.   72

141 **Hermit's Cave,** Eddy Co., NM.   Est. 1600 m, 5021 m ee.
12,900 ± 350 on charcoal from hearth (W-495); 11,850 ± 350 on
log from extinct mammal horizons (W-498); 12,270 ± 450 on log
from mouth of cave (W-499) (Hester, 1960).   Ferdon, 1946;
Nowak, 1979; Schultz et al., 1970.
†Canis dirus, C. lupus, †Equus sp., †Mammuthus columbi, Marmota sp., †Navahoceros fricki, Notiosorex crawfordi, Sorex
monticolus/vagrans, S. nanus

142 **Hord Rock Shelter,** nr. Alpine, Brewster Co., TX.   Smith,
1934.   8, 72, 91

143 **Horned Owl Cave,** Albany Co., WY.   2439 m, 6880 m ee.
Mixed provenience, Late Pleistocene/Holocene.   Guilday et
al., 1967.
Antilocapra americana, Bos or Bison, †Camelops cf hesternus,
Cynomys cf leucurus, †Equus sp. (large), Lagurus curtatus,
Lepus townsendi or L. californicus, Marmota flaviventris,
Mustela frenata, M. vison, Neotoma cinerea, Ochotona princeps, Odocoileus cf hemionus, Ondatra zibethicus, Oreamnos cf
americanus, Ovis canadensis, Peromyscus sp., Phenacomys
intermedius, Spermophilus lateralis, Spermophilus sp., Sylvilagus sp., Tamias sp., Taxidea cf taxus, Vulpes vulpes

144 **Horsethief Draw,** Cochise Co., AZ.   Lindsay and Tessman,
1974.   72

145 **Howell's Ridge Cave,** UTEP Loc. 32, Grant Co., NM.   1675
m, 5099 m ee.   See Table 5 for dates.   Wisconsinan/Holocene.
Harris, 1977b; Van Devender and Worthington, 1977; UTEP.

? Bison, †Camelidae, Centrocercus urophasianus, Cryptotis parva, Cynomys ludovicianus, Dipodomys sp., Dipodomys spectabilis, Eptesicus fuscus, †Equus cf conversidens, Felis cf concolor, Homo sapiens, Lagurus curtatus, Lepus townsendi, †Mammuthus sp., Microtus mexicanus, M. montanus, M. pennsylvanicus, Neotoma albigula, N. cinerea, Notiosorex crawfordi, cf Oreortyx pictus, Perognathus sp. (large), Perognathus sp. (small), Reithrodontomys sp., Sigmodon sp., Spermophilus sp., Spilogale putorius, Sylvilagus auduboni or S. floridanus, S. nuttalli, Thomomys bottae, Urocyon or Vulpes, Ursus americanus

146 **Hudson Hope**, B.C.   Pre- or mid-Wisconsinan.   Churcher and Wilson, 1979.   133

147 **Hueco Tanks No. 1**, El Paso or Hudspeth Co., TX.   Ca. 1420 m, 4844 m ee.   13,500 ± 250 (A-1624).   Van Devender and Riskind, 1979.
Microtus sp.

148 **Huerfano Co.**, CO.   Kurten and Anderson, 1980.   28

149 **Human Corridor**, UTEP Loc. 31, Dry Cave, Eddy Co., NM. 1280 m, 4758 m ee.   <15,000, >12,000.   Harris, 1977b; UTEP. †Camelops cf hesternus, †Capromeryx sp., Dipodomys cf ordi, D. spectabilis, †Equus conversidens, †E. cf occidentalis, †Hemiauchenia sp., Lepus sp., Lynx rufus, cf Mephitis sp., Microtus cf mexicana, Mustela frenata, Neotoma cinerea, Peromyscus sp., Sigmodon sp., Sylvilagus nuttalli, Thomomys bottae, T. talpoides, Ursus americanus, Vulpes cf velox

150 **Hurley**, Cochise Co., AZ.   1273 m.   21,210 ± 770 (AS-988)

from †Mammuthus bone apatite--believed contaminated (from
unit dated 29,000 ± 2000 [A-396A] at Murray Springs).
Lindsay and Tessman, 1974. 72, 133

151 Independence Rock Gravel Pit, Natrona Co., WY. Anderson,
1974. 133

152 Indian Creek, ? Co., WY. Anderson, 1974. 133

153 Isleta Cave No. 1, UTEP Loc. 41, Bernalillo Co., NM.
1716 m, 5461 m ee. Includes Holocene material. Findley et
al., 1975; Harris and Findley, 1964; UTEP.
Antilocapra americana, Antrozous pallidus, †Arctodus sp., Bos
or Bison, †Camelops cf hesternus, Canis latrans, C. cf lupus,
cf Centrocercus urophasianus, Cynomys gunnisoni, Dipodomys
ordi, D. spectabilis, †Equus cf niobrarensis, Erethizon dorsatum, Lagurus curtatus, Lepus townsendi, Lynx rufus, †Mam-
muthus sp., Mephitis mephitis, Microtus cf pennsylvanicus,
Myotis thysanodes, Neotoma albigula, N. cinerea, N. ? micropus, Onychomys leucogaster, O. torridus, Ovis aries, †Panthe-
ra leo atrox, Perognathus cf flavus, Peromyscus sp., Spermophilus spilosoma, Sylvilagus auduboni, S. floridanus, Thomomys bottae, Ursus americanus, Vulpes macrotis, V. velox, V.
vulpes

154 Isleta Cave No. 2, UTEP Loc. 46, Bernalillo Co., NM.
1716 m, 5461 m ee. Includes Holocene material. Findley et
al., 1975; Harris and Findley, 1964; UTEP.
cf Antilocapra americana, †Arctodus sp., Brachylagus idahoensis, Canis latrans, C. lupus, cf Centrocercus urophasianus,
Cynomys gunnisoni, Dipodomys ordi or D. merriami, D. spectabilis, †Equus cf conversidens, †E. cf niobrarensis, †Hemi-

auchenia cf macrocephala, Homo sapiens, Lagurus curtatus,
Lepus californicus, L. townsendi, Lynx rufus, Marmota flaviventris, Mephitis cf mephitis, Mustela nigripes, M. cf vison,
Neotoma albigula, N. cinerea, N. ? micropus, Onychomys leucogaster, O. torridus, Ovis aries, Perognathus flavus, P.
intermedius, Peromyscus spp., Reithrodontomys sp., Sylvilagus
auduboni, S. floridanus, S. nuttalli, Taxidea taxus, Urocyon
cinereoargenteus, Ursus americanus, Vulpes macrotis, V.
velox, V. vulpes

155 **Jaguar Cave**, Lemhi Co., ID.   2255 m, 7017 m ee.   11,580 ±
250 (GX-395); 10,370 ± 350 on charcoal from ca. midway in the
deposits. Guilday and Adam, 1967.
Antilocapra americana, Bison sp., Brachylagus idahoensis,
†Camelops cf hesternus, †Canis dirus, C. familiaris, C.
latrans, C. cf lupus, Castor canadensis, Centrocercus urophasianus, Cervus elaphus, Dicrostonyx cf torquatus, †Equus conversidens, †Equus sp. (large), Erethizon dorsatum, Felis concolor, Gulo gulo, Lagurus cf curtatus, Lepus americanus, L.
cf townsendi, Lynx canadensis, L. rufus, Marmota cf flaviventris, †Martes nobilis, Mephitis mephitis, Microtus cf montanus, Mustela frenata, M. nigripes, Neotoma cinerea, Ochotona
princeps, Odocoileus sp., Ovis canadensis, †Panthera leo
atrox, Peromyscus sp., Rangifer tarandus, Spermophilus cf
lateralis, S. richardsoni, S. townsendi, Spilogale putorius,
Sylvilagus cf nuttalli, Tamias sp., Taxidea taxus, Thomomys
cf talpoides, Ursus arctos, Vulpes cf macrotis, V. vulpes

156 **Jean D'or Prairie**, Alberta. Churcher and Wilson, 1979.
39

157 **Jimenez Cave**, UTEP Loc. 91, Chihuahua, Mexico. 1450 m,

4339 m ee.   Henry Messing, per. comm.; UTEP.
Antilocapra americana, Antrozous pallidus, Canis latrans,
†Capromeryx sp., Cryptotis parva, Cynomys sp., Dipodomys
spectabilis or D. nelsoni, †Equus sp., Felis concolor, Geococcyx californianus, †Leporidae, Lepus californicus, Lynx
rufus, Microtus pennsylvanicus, Mustela nigripes, Myotis sp.,
Neotoma albigula, N. ? cinerea, N. cf floridana, N. lepida,
N. micropus, Notiosorex crawfordi, cf Oreortyx pictus, Pappogeomys castanops, Peromyscus sp., Reithrodontomys fulvescens,
Sigmodon sp., Spermophilus cf spilosoma, S. variegatus, Sylvilagus auduboni, Tadarida sp., Thomomys cf umbrinus, Ursus
americanus

158 Joseph City, Navajo Co., AZ. Lindsay and Tessman, 1974.
72

159 Kassler Quadrangle, CO. ^{14}C date on mammoth tooth of
10,200 ± 350 (W-401) considered contaminated by Haynes
(1964). 133

160 Keams Canyon, Navajo Co., AZ. Lindsay and Tessman, 1974.
11, 17, 26, 72, 133, 194

161 Klamath River, Siskiyou Co., CA. Kurten and Anderson,
1980. 92

162 Kokoweef Cave, San Bernardino Co., CA. Pers. comm.,
Robert Reynolds via George Jefferson. cf 4, 8, 15, ? 24, 26,
cf 31, 34, ? 40, 63, cf 76, 92, 107, 111, 117, cf 127, cf
138, 150, 158, 165, 181, cf 183, 191, 193, 194, 214, 220,
231, 248, 265, cf 294, cf 297, cf 301, 304, cf 309, 323, 334,
338, 350, cf 356

163 La Mirada, Los Angeles and Orange counties, CA. <?100 m, 3653 m ee. 8550 ± 100 (UCLA-1321) on wood; Pleistocene mammals immediately above and below. Miller, 1971. †Camelops cf hesternus, †Canis cf dirus, C. cf latrans, †Equus sp., Lynx cf rufus, †Mammut americanum, †Megalonyx sp., Microtus cf californicus, Odocoileus cf hemionus, Peromyscus sp., Sylvilagus sp., Thomomys bottae, Urocyon cf cinereoargenteus, Ursus americanus

164 Lake Lahontan, NV. Nowak, 1979; Russell, 1885. 17, 35, cf 82, ? 134, 216

165 Lance Creek nr. junction with Cheyenne River, Niobrara Co., WY. Walker and Frison, 1980. 133

166 Laramie, Albany Co., WY. Walker and Frison, 1980. 76, 133

167 Laramie and North Platte river terraces, SE Quad. WY. Mears, 1981. 8, 22, 27, 76, 133, 210, 216

168 Lea Co., NM. McDonald, 1981. 21

169 Lehner Ranch, Cochise Co., AZ. 1277 m, 4648 m ee. Six ^{14}C dates average 11,260 ± 360 for Clovis cultural level, with 10,410 ± 190 (A-33) above that level and 11,600 ± 400 (A-478b) below (Haynes, 1964). Lindsay and Tessman, 1974. 17, 72, 131, 134, 216, 331

170 Leikum, Cochise Co., AZ. 1382 m. Kurten and Anderson, 1980. 134

171 Lewis Hill, Cochise Co., AZ. 1311 m. Lindsay and
Tessman, 1974. 26, 72

172 Lindenmeier, Larimer Co. ?, CO. 10,850 ± 550 (I-141)
(Frison et al., 1978). Folsom level date of 10,780 ± 135
(I-141) according to Haynes (1964). Frison et al., 1978;
McDonald, 1981. 21, 26

173 Lindsey Ranch, Cochise Co., AZ. 1251 m. Lindsay and
Tessman, 1974. 17, 72, 133

174 Lingle, Goshen Co., WY. Anderson, 1974. 133

175 Little Box Elder Cave, Converse Co., WY. 1676 m, 6224 m
ee. Late Wisconsinan and Holocene. Anderson, 1968; Mears,
1981; Walker, 1982b.
? Alces alces, Antilocapra americana, †Arctodus simus, Bison
bison, †Camelops cf hesternus, Canis latrans, C. lupus, Castor canadensis, Cervus elaphus, Clethrionomys cf gapperi,
Cryptotis sp., Cynomys sp., Dicrostonyx cf torquatus, Eptesicus fuscus, Equus caballus, †E. cf conversidens, Erethizon
dorsatum, Felis concolor, Gulo gulo, ? †Hemiauchenia sp.,
Homo sapiens, Lagurus curtatus, Lasionycteris noctivagans,
Lepus cf townsendi, Lynx rufus, Marmota flaviventris, †Martes
nobilis, Mephitis mephitis, Microtus longicaudus, M. montanus, M. pennsylvanicus, Mustela frenata, M. nigripes, Myotis
evotis, M. leibii, M. thysanodes, M. volans, †Navahoceros
fricki, Neotoma cinerea, Ochotona princeps, Odocoileus hemionus, Ondatra zibethicus, Oreamnos americanus, Ovis canadensis, †Panthera leo atrox, Perognathus sp., Peromyscus sp.,
Phenacomys cf intermedius, Pitymys ochrogaster, Sorex hoyi,
S. palustris, Sorex sp., Spermophilus lateralis, S. tride-

cemlineatus, S. variegatus, Spilogale putorius, Sylvilagus
sp., †Symbos sp., Tamias cf minimus, Taxidea taxus, Thomomys
sp., Ursus arctos, Vulpes vulpes

176 Little Canyon Creek Cave, Big Horn Co. ?, WY. Est. 1430
m, 6138 m ee. Symbos level lies below stratigraphic uncon-
formity ^{14}C-dated at 10,170 ± 250 (RL-641). Frison and Walker, 1978; Mears, 1981; Walker, 1982b.
†Acinonyx trumani, †Canis cf dirus, Clethrionomys gapperi,
Dicrostonyx torquatus, Lagurus curtatus, †Martes nobilis,
Microtus pennsylvanicus, Mustela nigripes, Ochotona princeps,
Oreamnos americanus, †Ovis cf canadensis catclawensis, Phenacomys cf intermedius, †Symbos sp.

177 Livermore, Alameda Co., CA. Harington, 1969; Nowak,
1979. 21, 27, 31, 216

178 Loon River, Alberta. Churcher and Wilson, 1979. 27, 136

179 Lopez, Pima Co., AZ. Lindsay and Tessman, 1974. 133

180 Los Angeles Basin, exclusive of Rancho La Brea, Costeau
Pit, Newport Bay Mesa Localities 1066 and 1067, San Pedro
Loc. UCMP V-2047, and La Mirada, Los Angeles Co., CA. <100
m. Miller, 1971; Nowak, 1979; Stock, 1944. 8, 19, cf 20, cf
21, 26, 31, cf 34, 36, 72, 82, 105, 107, 131, cf 134, 142, cf
160, 183, cf 195, 215, cf 216, 251, cf 277, 291, 331, cf 339

181 Lost Valley, UTEP Loc. 1, 17, Dry Cave, Eddy Co., NM.
1280 m, 4758 m ee. 29,290 ± 1060 (TX-1774) on bone carbonates. Harris, 1977b; UTEP.
†Camelops cf hesternus, Canis latrans, †Capromeryx sp., Cyno-

mys ludovicianus, Dipodomys sp., D. spectabilis, †Equus sp.,
cf Felis concolor, †Leporidae, Lepus californicus, Lynx rufus, Mustela frenata, Neotoma albigula, N. micropus, †Neotoma
Species A, †Neotoma Species B, Onychomys leucogaster, Pappogeomys cf castanops, Perognathus sp., Peromyscus spp., Pitymys ochrogaster, Plecotus sp., Sigmodon sp., cf Sorex sp.,
Spermophilus sp., Spilogale putorius, Sylvilagus nuttalli,
Tadarida brasiliensis, T. macrotis, ? Tamias sp., †Tapirus
sp., Urocyon cinereoargenteus

182 **Lower Sloth Cave**, Culberson Co., TX.   2000 m, 5424 m ee.
11,590 ± 230 (A-1519).   Logan, 1983.
Antrozous pallidus, cf Bassariscus astutus, †Canis dirus or
C. lupus, Cryptotis parva, Cynomys cf gunnisoni, Eptesicus
fuscus, Erethizon dorsatum, Marmota flaviventris, Microtus
mexicanus, Mustela frenata, Myotis leibii, M. thysanodes, M.
velifer, Neotoma albigula, N. cinerea, N. mexicana, N. micropus, †Nothrotheriops shastensis, Notiosorex crawfordi, Odocoileus sp., Onychomys leucogaster, O. torridus, Ovis canadensis, Peromyscus sp., Plecotus townsendi, Sciurus sp., Sorex
cinereus, S. monticolus/vagrans, Spermophilus spilosoma, S.
variegatus, Spilogale putorius, Sylvilagus sp., Tadarida brasiliensis, Tamias sp., Tamiasciurus hudsonicus, cf Thomomys
bottae

183 **Lucy Site**, Torrance Co., NM.   Jelinek, 1957.   133

184 **Mahogany Buttes**, Washakie Co., WY.   Anderson, 1974.   133

185 **Manix Lake**, San Bernardino Co., CA.   Buwalda, 1914; Kurten and Anderson, 1980.   6, 25, cf 29, 73, 74, 132, 188, 212

186 **Manzano Cave**, Torrance Co., NM. 2745 m. Hibben, 1941.
26, 188

187 **Maricopa**, Kern Co., CA. 260 m, 4005 m ee. 13,860 BP
(Nowak, 1979). Kurten and Anderson, 1980; McDonald, 1967;
Nowak, 1979.
Antilocapra americana, †Canis dirus, C. latrans, C. lupus,
†Equus sp., Leporidae, Microtus sp., Panthera sp. (giant
jaguar), †Smilodon sp., Taxidea taxus

188 **McCammon**, Bannock Co., ID. McDonald, 1981. 19, 21

189 **McCullum Ranch**, Roosevelt Co., NM. 15,770 ± 760 (A-375).
Haynes and Agogino, 1966. 17, 25, 69, 72, 133

190 **McKittrick**, Kern Co., CA. 320 m, 4119 m ee. Schultz,
1938.
Ammospermophilus cf nelsoni, Antilocapra americana, Antrozous
pallidus, †Arctodus simus, †Bison antiquus, †Camelops hesternus, †Canis dirus, C. latrans, C. lupus, †Capromeryx minor,
Cervus sp., Dipodomys cf ingens, †Equus occidentalis, †Euce-
ratherium collinum, Geococcyx californianus, †Glossotherium
harlani, †Hemiauchenia macrocephala, Lepus californicus,
†Mammut americanum, †Mammuthus columbi, ? †Megalonyx sp.,
Mephitis mephitis, Microtus californicus, Mustela frenata,
Neotoma fuscipes, N. lepida, Odocoileus sp., ? Onychomys sp.,
Panthera leo atrox, Perognathus cf inornatus, Peromyscus cf
californicus, †Platygonus cf compressus, ? Reithrodontomys
sp., Sorex cf ornatus, Spermophilus cf variegatus, Spilogale
putorius, Sylvilagus auduboni, S. bachmani, Taxidea taxus,
Thomomys bottae, Ursus americanus, Vulpes macrotis

191 **Medicine Hat Fauna 2**, Alberta. 11,000 BP suggested.
Harington, 1978. 17, ? 25, 35, 76, 109, 132, 136

192 **Medicine Hat Fauna 3**, Alberta. Late Wisconsinan. 15,000
BP felt too early or too late in view of ice front positions.
Harington, 1978. 26, 35, 76, 135

193 **Medicine Hat Fauna 4**, Alberta. Late mid-Wisconsinan.
Harington, 1978.
? Antilocapra cf americana, †Camelops hesternus, †Canis cf
dirus, †Equus conversidens, †Mammuthus sp., †Nothrotheriops
shastensis, Odocoileus sp., †Smilodon fatalis

194 **Medicine Hat Fauna 5**, Alberta. Early mid-Wisconsinan.
Harington, 1978.
? Antilocapra cf americana, †Bison sp. (large), †Camelops
hesternus, Canis cf latrans, Cynomys cf ludovicianus, †Equus
conversidens, †E. cf giganteus, †Hemiauchenia ? macrocephala,
Lepus cf townsendi, †Mammuthus sp., Microtus sp., Odocoileus
sp., †Smilodon fatalis, Spermophilus richardsoni, Spilogale
putorius, Thomomys cf talpoides

195 **Medicine Hat Fauna 6**, Alberta. Early Wisconsinan.
Harington, 1978. cf 27, 76, 136

196 **Medicine Hat Fauna 7**, Alberta. Sangamonian. Harington,
1978.
? Alces sp., ? Antilocapra cf americana, †Bison cf latifrons,
†Camelops hesternus, †Canis dirus, C. lupus, Cervus cf elaphus, Cynomys cf ludovicianus, Erethizon dorsatum, †Equus
(Amerhippus) sp., †E. conversidens, †E. scotti, †Hemiauchenia
sp., Lepus cf townsendi, Lynx canadensis, †Mammuthus columbi,

†Megalonyx sp., Microtus sp., Mustela (Putorius) cf nigripes, Odocoileus sp., Ondatra zibethicus, Ovis canadensis, †Panthera leo atrox, Procyon lotor, Rangifer sp. (small), R. tarandus, Spermophilus richardsoni, Sylvilagus floridanus, Thomomys cf talpoides, Vulpes vulpes

197 **Mercers Cave**, Calaveras Co., CA.   143

198 **Mercury Ridge**, Clark Co., NV.   1250 m.   12,700 ± 200. Merhinger and Ferguson, 1969.   138

199 **Mesa De Maya**, Las Animas Co., CO.   1800 m, 5759 m ee. Sangamonian.   Hager, 1974.
Bison sp., †Camelops hesternus, Canis latrans, Cynomys gunnisoni, C. ludovicianus, †Equus conversidens, †Equus sp. (large), Geomys bursarius, †Mammuthus cf imperator, Neotoma sp., Perognathus hispidus, †Peromyscus cragini, †P. progressus, Pitymys ochrogaster, Spermophilus cf tridecemlineatus, Sylvilagus sp.

200 **Mescal Cave**, San Bernardino Co., CA.   1550 m, 5349 m ee. Brattstrom, 1958; Mehringer and Ferguson, 1969.
Marmota flaviventris, Microtus sp., Neotoma cf cinerea, Ochotona princeps, Spermophilus lateralis

201 **Milan Site** nr. Three Hills, Alberta.   [14]C dates of 9630 ± 300 (GSC-1894) and 9670 ± 160 (I-8579), but what appears to be a glacial till overlies.   Harington, 1978.   21, 45

202 **Mineral Hill Cave**, Elko Co., NV.   Late Wisconsinan and Holocene.   McGuire, 1980.   8, 34, 72, 91, 107, 117, 127, 137,

149, 158, 175, 214, 231, 289, 303, 308, 323, 334, 338, 350, 355

203 **Minidoka,** Minidoka Co., ID. McDonald, 1981. 23

204 **Minika River** nr. Trutch, B.C. Churcher and Wilson, 1979. 131

205 **Mockingbird Gap,** Socorro Co., NM. Kurten and Anderson, 1980. 134

206 **Monolith Gravel Quarry,** Albany Co., WY. Anderson, 1974. 27, 72, 216

207 **Mosan Wash,** Cochise Co., AZ. 1234 m. Lindsay and Tessman, 1974. 133

208 **Muav Caves,** Mohave Co., AZ. Surface: 11,140 ± 160 (A-1212) and 11,290 ± 170 (A-1213). Long and Martin, 1974. 189

209 **Murpheys,** Calaveras Co., CA. Pleistocene. Nowak, 1979. 34

210 **Murray Springs Arroyo Fauna,** Cochise Co., AZ. 1273 m, 4644 m ee. Possibly earlier than Sangamonian. Lindsay, 1978. 31, 34, 63, 149, 175, 200, 231, 262, 274, 289, 308, 338

211 **Murray Springs Site, M.S. Conduit,** Cochise Co., AZ. 1273 m, 4644 m ee. Lindsay and Tessman, 1974. 17, 26, 72, 133

212 **Murray Springs Site, Occupation Level** (UALP 63), Cochise Co., AZ. 1273 m, 4644 m ee. Lindsay and Tessman, 1974. 17, 26, 72, 117, 133, 149, 175, 220

213 **Murray Srings Site, Portell Conduit** (UALP 7024), Cochise Co., AZ. 1273 m, 4644 m ee. Lindsay and Tessman, 1974. 17, 26, 72, 95, 112, 133, 331

214 **Murray Springs Site, Unit D** (UALP 6954), Cochise Co., AZ. 1273 m, 4644 m ee. 29,000 ± 2000 (A-896). Haynes, 1968, Lindsay and Tessman, 1974. 17, 26, 72, 134

215 **Murray Springs Site, Unit E** (UALP 6955), Cochise Co., AZ. 1273, 4644 m ee. 21,200 ± 500 (A-897). Haynes, 1968; Lindsay and Tessman, 1974. 17, 26, 72, 133

216 **Muskox Cave**, Eddy Co., NM. 1600 m, 5024 m ee. 25,500 ± 1100 to 18,140 ± 200 for much of deposit. Logan, 1981. †Acinonyx trumani, cf Antilocapra americana, Antrozous pallidus, Bassariscus astutus, cf †Camelops, †Canis dirus, Canis sp. (wolf), Conepatus sp., Cryptotis parva, Eptesicus fuscus, †Equus sp., Erethizon dorsatum, †Euceratherium collinum, Felis concolor, Lynx rufus, Marmota flaviventris, Microtus mexicanus, M. pennsylvanicus, Mustela frenata, Myotis thysanodes, M. velifer, Neotoma albigula, N. cinerea, N. mexicana, N. micropus, Notiosorex crawfordi, Ondatra zibethicus, Onychomys leucogaster, O. torridus, †Oreamnos-like bovid, Ovis canadensis, †Panthera leo atrox, Perognathus flavus, Peromyscus sp., Pitymys ochrogaster, Plecotus townsendi, Reithrodontomys cf fulvescens, Sorex cinereus, S. merriami, S. monticolus/vagrans, S. palustris, Spermophilus variegatus, Spilogale putorius, †Stockoceros conklingi, †S. onusrosagris, Sylvila-

gus cf floridanus, S. cf nuttalli, Tadarida sp., Tamias sp., Tamiasciurus hudsonicus, Thomomys cf bottae

217 **Naco**, Cochise Co., AZ. 1381 m, 4752 m ee. Clovis date of 9250 ± 300 (A-9 + 10) by solid-carbon method believed too young; older on stratigraphic evidence (Haynes, 1964). Lance, 1953. 17, 72, 130, 134

218 **Natural Trap Cave**, Big Horn Co., WY. 1400 m, 6215 m ee. Stratum 2 ca. 11,000 to 14,000 BP; stratum 3 at 17,620 on horse collagen near top to 20,170 on sheep collagen at bottom. Martin and Gilbert, 1978; Mears, 1981.
†Acinonyx trumani, Antilocapra americana, †Arctodus simus, Bison bison, †Camelops cf hesternus, †Canis dirus, C. latrans, C. lupus, Cervidae (large), Dicrostonyx torquatus, †Equus (Amerhippus), †E. conversidens, †E. (Hemionus), Gulo gulo, Lagurus curtatus, Lepus sp., †Mammuthus sp., Marmota sp., †Martes nobilis, Microtus sp., Mustela sp., Neotoma sp., Ochotona princeps, Ovis canadensis, †Panthera leo atrox, Peromyscus sp., Spermophilus sp., Sylvilagus sp., †Symbos sp., Tamias sp., Tree Squirrel, Vulpes vulpes

219 **Navar Ranch No. 13**, El Paso Co., TX. Van Devender and Riskind, 1979. cf 155

220 **New Cave**, Eddy Co., NM. Lawrence, 1960. 321

221 **New La Bajada Hill**, Santa Fe Co., NM. 1800 m. Stearns, 1942. cf 22, 138, 231, 330, 338

222 **Newport Bay Mesa Loc. 1066**, Orange Co., CA. <100 m. Sangamonian. Miller, 1971.

†Bison cf latifrons, †Canis cf dirus, †Mammut or †Mammuthus, †Megalonyx cf jeffersoni, Neotoma sp., †Nothrotheriops shastensis, cf Perognathus sp., Spermophilus cf beecheyi, Sylvilagus cf auduboni, †Tapirus sp., Thomomys cf bottae

223 **Newport Bay Mesa Loc.** 1067, Orange Co., CA. <100 m. Miller, 1971. 9, cf 64, 72, 107, cf 129, 150, 181, 191, cf 221, cf 237, 241, 265, 267, 284, cf 303, 310, 339

224 **Nichols Site,** Maricopa Co., AZ. Lindsay and Tessman, 1974. 72, 133, 188

225 **Orr Cave,** Beaverhead Co., MT. Kurten and Anderson, 1980. 161

226 **Paisley Cave** (=Five Mile Point Cave), OR. 7610 ± 120 on charcoal above Camelops and Equus. Hester, 1960. 26, 72

227 **Palomas Creek Cave,** Sierra Co., NM. UTEP. 72, 149, 175, 214, 308, 338, cf 357

228 **Papago Springs Cave,** Santa Cruz Co., AZ. 1586 m, 4957 m ee. Late Wisconsinan. Lindsay, 1978; Rea, 1980; Skinner, 1942.
Antrozous pallidus, †Bassariscus sonoitensis, Bison sp., †Camelidae, Canis latrans, C. lupus, Cervus sp., †Equus conversidens, †E. tau, Lepus sp., †Mammuthus sp., Marmota flaviventris, Mephitis mephitis, Microtus mexicanus, Myotis ? evotis, M. ? thysanodes, M. ? velifer, Neotoma ? mexicana or N. albigula, Onychomys ? leucogaster, Perognathus ? flavescens, Peromyscus ? boylii or P. truei, P. maniculatus, †Platygonus alemani, Plecotus ? townsendi, Spermophilus variegatus,

Spilogale putorius, †Stockoceros onusrosagris, Sylvilagus
auduboni, Tadarida brasiliensis, Tamias ? dorsalis, Taxidea
taxus, Thomomys ? bottae or T. umbrinus, Urocyon cinereoargenteus, Ursus americanus

229 Park City, Summit Co., UT. McDonald, 1981. 23

230 Parker, Douglas Co., CO. McDonald, 1981. 19, 76

231 Parsnip River, B.C. Kurten and Anderson, 1980. 214

232 Peace River Bridge, Alberta. Churcher and Wilson, 1979.
133

233 Picacho, Dona Ana Co., NM. UTEP. cf 63

234 Pine Springs, WY. 11,830 ± 410 (GXU-355). Frison et
al., 1978. 26

235 Pirtleville, Cochise Co., AZ. Lindsay and Tessman, 1974.
133

236 Pit N and W of Animal Fair, UTEP Loc. 122, Dry Cave, Eddy
Co., NM. 1280 m, 4758 m ee. ?Interstadial. UTEP.
Canis lupus, †Capromeryx sp., Conepatus mesoleucus, Cynomys
cf ludovicianus, †Equus conversidens, †E. niobrarensis,
Lepus ? californicus, Microtus mexicanus, Neotoma cinerea,
Onychomys leucogaster, Panthera cf onca, Pitymys cf ochro-
gaster, Spilogale putorius, Sylvilagus auduboni, S. nuttalli,
Vulpes sp.

237 Pomerene West, Cochise Co., AZ. Lindsay and Tessman,

1974. 133

238 **Ponoka,** Alberta. Harington, 1978. 209

239 **Portage Pass,** B.C. >11,600 (I-2244A). Harington, 1978.
136

240 **Potter Creek Cave,** Shasta Co., CA. 457 m, 4844 m ee.
Hutchison, 1967; Kellogg, 1912; Nowak, 1979.
Antrozous pallidus, Aplodontia rufa, †Arctodus simus, Bassariscus astutus, Bison sp., †Canis dirus, C. lupus, †Desmodus
stocki, †Equus occidentalis, †E. pacificus, †Euceratherium
collinum, Felis cf colcolor, Felis sp., Glaucomys sabrinus,
Lepus americanus, L. californicus, Lynx rufus, †Mammut americanum, †Mammuthus primigenius, Marmota flaviventris, †Martes
nobilis, †Megalonyx ? jeffersoni, Mephitis mephitis, Microtus
californicus, Mustela frenata, Neotoma cinerea, †Nothrotheri-
ops shastensis, Odocoileus sp. A, Odocoileus sp. B, Oreamnos
americanus, Oreortyx pictus, ? †Panthera leo atrox, ? †Platy-
gonus sp., Scapanus ? latimanus, Spermophilus beecheyi, S.
lateralis, Spilogale putorius, Sylvilagus auduboni, Tamias
sp., Tamiasciurus douglasi, Taxidea taxus, Thomomys bottae,
†T. microdon, Urocyon cinereoargenteus, Ursus americanus,
Vulpes vulpes

241 **Powder River,** Johnson Co. ?, WY. Anderson, 1974. 135

242 **Prospects Shelter,** Big Horn Co., WY. Ca. 1400 m, 6215 m
ee. 16,272 ± 8.5% thermoluminescent (MATL-29); 10,005 ±
14.9% (MATL) immediately antedates last Dicrostonyx record.
Chomko, 1978; Mears, 1981; Walker, 1982b.
Antilocapra americana, †Arctodus simus, Bison bison, †Camel-

ops cf hesternus, Canis lupus, Dicrostonyx torquatus, †Equus
sp., Lagurus curtatus, Microtus pennsylvanicus, Ochotona
princeps, †Ovis cf canadensis catclawensis, Phenacomys cf
intermedius, †Symbos sp.

243 **Providence Mts.**, San Bernardino Co., CA.   Wilson, 1942.   138

244 **Pyeatt Cave**, Cochise Co, AZ.   Lindsay and Tessman, 1974.
95, 189, 308

245 **Quarai**, Torrance Co., NM.   Hibben, 1941.   133

246 **Rainbow Beach**, Power Co., ID.   1335 m, 5936 m ee.   21,500
± 700 (WSU-1423), 31,300 ± 2300 (WSU-1424), both on collagen.
McDonald and Anderson, 1975.
Antilocapridae, †Arctodus simus, †Bison antiquus, †B. latifrons, Brachylagus idahoensis, †Camelops cf hesternus, †Canis
dirus, C. latrans, Castor canadensis, Centrocercus urophasianus, Cervidae, Cricetinae, Cynomys cf leucurus, †Equus sp.,
†Glossotherium harlani, Lepus sp., †Mammuthus sp., †Megalonyx
jeffersoni, Microtus cf montanus, Ondatra zibethicus, †Smilo-
don sp., Sorex sp., Spermophilus richardsoni, Sylvilagus sp.,
Thomomys townsendi

247 **Rampart Cave**, Mohave Co., AZ.   525 m, 4377 m ee.   A
series of dates from >40,000 (A-1042) to 10,780 ± 200
(A-1067) plus two rejected by Long and Martin (1974) of
10,400 ± 275 (I-442) and 10,035 ± 250 (L-473A).   Average surface date without these two, 11,070.   Harington, 1972; Van
Devender et al., 1977; Wilson, 1942.
Bassariscus astutus, †Equus sp., Erethizon dorsatum, Felis
concolor, Lepus cf californicus, Lynx sp., Marmota flaviven-

tris, Neotoma lepida, N. cf mexicana, N. stephensi, †Nothro-
theriops shastensis, †Oreamnos harringtoni, Ovis canadensis,
Peromyscus sp., Spermophilus sp.

248 **Rancho La Brea**, Los Angeles Co., CA. Ca. 15 m, 3653 m
ee. Akersten et al., 1979; Miller, 1968; Stock, 1956.
Antilocapra americana, †Arctodus simus, Bassariscus astutus,
†Bison antiquus, †B. latifrons, †Camelops hesternus, †Canis
dirus, C. cf familiaris, C. latrans, C. lupus, †Capromeryx
minor, Dipodomys agilis, †Equus occidentalis, Felis concolor,
Geococcyx californianus, †Glossotherium harlani, †Hemiauche-
nia sp., Lasiurus cinereus, Lepus californicus, Lynx rufus,
†Mammut americanum, †Mammuthus columbi, †Megalonyx jeffersoni, Mephitis mephitis, Microtus californicus, Mustela frenata, Neotoma sp., †Nothrotheriops shastensis, Notiosorex
crawfordi, Odocoileus sp., Onychomys torridus, †Panthera leo
atrox, Perognathus californicus, Peromsycus maniculatus,
†Platygonus sp., Procyon lotor, Reithrodontomys megalotis,
Scapanus latimanus, †Smilodon fatalis, Sorex cf ornatus,
Spermophilus beecheyi, Spilogale putorius, Sylvilagus auduboni, S. bachmani, †Tapirus sp., Taxidea taxus, Thomomys bottae, Urocyon cinereoargenteus, Ursus americanus, Ursus arctos

249 **Rawhide Butte Area**, Goshen and/or Niobrara counties, WY.
>10,500 ± 350. Anderson, 1974. 133

250 **Rawlins**, Carbon Co., WY. Anderson, 1974. 27

251 **Richville Gravels**, Apache Co., AZ. Lindsay and Tessman,
1974. 72, 105, 133

252 **Robledo Cave**, Dona Ana Co., NM. UTEP. cf 147

253 Rodeo, Contra Costa Co., CA. Merritt, 1978. 235

254 Rome, Malheur Co., OR. Harington, 1969. 216

255 **Room of the Vanishing Floor,** UTEP Loc. 26, 27, Dry Cave,
Eddy Co., NM. 1280 m, 4758 m ee. 33,590 ± 1500 (TX-1773) on
bone carbonates. Harris, 1977<u>b</u>; UTEP.
†<u>Camelops</u> sp., <u>Canis</u> cf <u>latrans</u>, <u>C</u>. <u>lupus</u>, †<u>Capromeryx</u> sp.,
<u>Cynomys</u> sp., <u>Dipodomys</u> sp. (small), †Edentata, †<u>Equus</u> occi-
dentalis, †<u>Equus</u> sp., <u>Geococcyx</u> californianus, †Leporidae,
<u>Lepus</u> californicus, <u>Lynx</u> <u>rufus</u>, <u>Marmota</u> or <u>Erethizon</u>, <u>Myotis</u>
velifer, <u>Neotoma</u> <u>albigula</u>, †<u>Neotoma</u> Species A, †<u>Neotoma</u> Species B, <u>Panthera</u> <u>onca</u>, <u>Pappogeomys</u> ? castanops, <u>Perognathus</u>
(Chaetodipus) sp., <u>P</u>. (Perognathus) sp., <u>Peromyscus</u> cf <u>mani-</u>
culatus, <u>Peromyscus</u> sp., <u>Plecotus</u> sp., <u>Reithrodontomys</u> (Reithrodontomys) sp., <u>Spermophilus</u> (Ictidomys) sp., <u>S</u>. variegatus, <u>Sylvilagus</u> <u>auduboni</u>, <u>S</u>. <u>nuttalli</u>, cf <u>Taxidea</u> sp., Urocyon cinereoargenteus, cf <u>Ursus</u> sp.

256 **Rupert,** Minidoka Co., ID. McDonald, 1981. 23

257 **Sabertooth Camel Maze,** UTEP Loc. 2, 5, Dry Cave, Eddy
Co., NM. 1280 m, 4758 m ee. 25,160 ± 1730 (TX-1775) on bone
carbonates. Harris, 1977<u>b</u>; UTEP.
<u>Bassariscus</u> sp., †<u>Camelops</u> cf <u>hesternus</u>, <u>Canis</u> <u>lupus</u>, †<u>Capro-</u>
meryx sp., <u>Dipodomys</u> sp. (small), <u>Felis</u> sp., †Leporidae,
<u>Lepus</u> californicus, <u>Neotoma</u> <u>albigula</u>, <u>Sylvilagus</u> <u>auduboni</u>,
<u>Tadarida</u> <u>macrotis</u>, <u>Urocyon</u> cinereoargenteus, <u>Vulpes</u> cf <u>velox</u>

258 **Salt Creek,** UTEP Loc. 34, Reeves Co., TX. McDonald,
1981; UTEP. 21, cf 26, cf 34, cf 54, cf 76, cf 81, cf 134

259 Samwel Cave, Shasta Co., CA. 455 m, 4842 m ee. Arata and Hutchison, 1964; Graham, 1959; Stock, 1918. Aplodontia rufa, †Canis dirus, C. latrans, C. lupus, Castor canadensis, Erethizon dorsatum, †Equus occidentalis, †Euceratherium collinum, Felis cf concolor, Glaucomys sabrinus, Lepus americanus, †Mammuthus sp., †Martes nobilis, †Megalonyx sp., Mephitis mephitis, Microtus californicus, Mustela frenata, Mustela sp., Neotoma cinerea, †Nothrotheriops shastensis, Odocoileus sp. A, Odocoileus sp. B, Oreamnos sp., Oreortyx pictus, Peromyscus maniculatus, Procyon lotor, Sciurus griseus, Spermophilus beecheyi, S. lateralis, Sylvilagus auduboni, Tamiasciurus douglasi, Thomomys bottae, †T. microdon, Urocyon cinereoargenteus, Ursus americanus, Vulpes sp.

260 San Francisco Bay Area, several counties, CA. McDonald, 1981; Savage, 1951. 6, 21, 23, 27, 31, cf 37, 72, cf 105, 131, cf 134, 142, 289, 331

261 San Josecito Cave, Nuevo Leon, Mexico. 2250 m, 4818 m ee. Howard, 1971; Jakway, 1958; Jones, 1958; Kurten, 1973; Nowak, 1979; Russell, 1968. 16, 25, 31, 34, 35, ? 47, cf 49, 51, 53, 60, cf 71, 76, 92, 97, 98, 99, 114, 115, 122, 127, 137, 146, 154, 162, 174, 176, 189, 205, 208, 216, 217, 218, 219, 233, 256, 265, ? 270, 275, 277, 279, 286, cf 296, 304, 306, 308, 312, 318, 334, 345, 347, 353

262 San Pedro, Los Angeles Co., CA. <100 m, 3653 m ee. Sangamonian. Lyon, 1938; Miller, 1971; Schultz, 1938. †Bison cf latifrons, †Camelops cf hesternus, †Canis cf dirus, †Capromeryx minor, †Equus sp., Felis cf concolor, Lepus sp., †Megalonyx jeffersoni, Microtus cf californicus, Neotoma cf fuscipes, †Nothrotheriops shastensis, Odocoileus cf hemionus,

†Panthera cf leo atrox, †Smilodon cf fatalis, Spermophilus beecheyi, Sylvilagus cf bachmani, Thomomys bottae

263 San Pedro Valley, Cochise Co., AZ. Lindsay, 1978. 26, 72, 133, 149

264 San Rafael Aq., Cochise Co., AZ. Lindsay and Tessman, 1974. 17, 26, 72

265 San Simon Sink, UTEP Loc. 81, Lea Co., NM. UTEP. cf 194

266 Sandia Cave, Sandoval Co., NM. 2210 m, 5955 m ee. Hibben, 1941. 21, 26, cf 35, 77, cf 82, 131, 133, 188

267 Santa Maria Oil Spring, Kern Co., CA. 82

268 Saskatoon, Saskatchewan. Sangamonian or cf. Wisconsinan interstadial. >34,000 (S-426). Harington, 1978.
Bison sp., †Camelops sp., Cervus elaphus, †Equus scotti, †Mammuthus columbi, Taxidea taxus

269 Schaldack, Cochise Co., AZ. 1273 m. Haynes, 1968. 72, 133

270 Schuiling Cave, San Bernardino Co., CA. 658 m, 4403 m ee. Downs et al., 1959; G. Jefferson, pers. comm.
†Camelidae ? †Camelops hesternus, Canis cf lupus, †Capromeryx minor, ? Dipodomys, †Equus sp. (large), †Equus sp. (small), Felis cf concolor, †Hemiauchenia sp., Homo sapiens, Lepus sp., Lynx sp., Neotoma sp., Ovis sp., Perognathus sp., ? Procyon sp., ? Spermophilus sp., Sylvilagus sp., Taxidea cf taxus, ? Thomomys sp., Urocyon sp.

271 **Secret Valley**, Lassen Co., CA. McDonald, 1981. 23

272 **Seff Loc.**, Cochise Co., AZ. 1122 m. Haynes, 1968. 25, 72, 133

273 **Selby**, Yuma Co., CO. 1173 m, 5453 m ee. Below Clovis level. >11,800, <29,000. Graham, 1981.
†Bison cf antiquus, †Camelops hesternus, †Equus sp., Felis sp., †Glossotherium harlani, †Mammuthus columbi, Microtus sp., Odocoileus sp., †Panthera cf leo atrox, †Platygonus compressus, Spermophilus richardsoni, S. spilosoma or S. tridecemlineatus, Thomomys talpoides

274 **Shawnee Creek**, Converse Co., WY. Walker and Frison, 1980. 133

275 **Sheaman Site**, Niobrara Co., WY. Ca 3920 m. Clovis age. Walker, 1982b; Walker and Frison, 1980. 9, 30, 133, 151, 157, 315

276 **Sheep Range, Flaherty Mesa No. 1**, Clark Co., NV. 1770 m, 5676 m ee. 20,380 ± 340 (WSU-1862). Thompson and Mead, 1982.
Ochotona cf princeps

277 **Sheep Range, Long Canyon Saddle Midden**, Clark Co., NV. 1800 m, 5706 m ee. 30,400 ± 1500 (A-1379). Thompson and Mead, 1982. 155, 175

278 **Sheep Range, Southcrest Midden**, Clark Co., NV. 1990 m, 5896 m ee. 21,700 ± 500 (LJ-2840). Thompson and Mead, 1983.
Microtus sp., Neotoma cf lepida, Sylvilagus sp.

279 **Shelter Cave,** Dona Ana Co., NM. 1435 m, 4859 m ee.
11,330 ± 370 on sloth dung (Van Devender and Spaulding,
1979). Harris, 1977<u>b</u>.
<u>Aegolius</u> <u>funereus</u>, †<u>Camelops</u> sp., <u>Canis</u> <u>latrans</u>, †<u>Capromeryx</u>
sp., <u>Centrocercus</u> <u>urophasianus</u>, †<u>Equus</u> sp. (large), †<u>Equus</u>
sp. (small), †<u>Geococcyx</u> <u>californianus</u> <u>conklingi</u>, ? Geomyidae,
<u>Lynx</u> <u>rufus</u>, <u>Mephitis</u> <u>mephitis</u>, <u>Microtus</u> cf <u>montanus</u>, <u>Neotoma</u>
<u>cinerea</u>, †<u>Nothrotheriops</u> sp., <u>Notiosorex</u> <u>crawfordi</u>, <u>Odocoi-</u>
<u>leus</u> sp., <u>Oreortyx</u> <u>pictus</u>, <u>Ovis</u> <u>canadensis</u>, <u>Spilogale</u> <u>puto-</u>
<u>rius</u>, †<u>Stockoceros</u> <u>conklingi</u>, <u>Taxidea</u> <u>taxus</u>, <u>Urocyon</u> or
<u>Vulpes</u>

280 **Sheridan Area,** Sheridan Co., WY. Anderson, 1974. 72, 133

281 **Shonto,** Navajo Co., AZ. Lindsay and Tessman, 1974. **6,**
26, 72, 93, 116, 133, 164, 175, 220, 231, 268, 338, 349

282 **Shoshone Falls,** Twin Falls Co., ID. Anderson and White,
1975. **261**

283 **Silver Bell Mts.,** Pima Co., AZ. Lindsay, 1978. **190**

284 **Silver Creek,** Summit Co., UT. 1950 m, 6284 m ee.
Sangamonian. Miller, 1976.
<u>Antilocapra</u> cf <u>americana</u>, †<u>Bison</u> ? <u>latifrons</u>, <u>Brachylagus</u> cf
<u>idahoensis</u>, †<u>Camelops</u> cf <u>hesternus</u>, †<u>Canis</u> cf <u>dirus</u>, <u>C</u>. ?
<u>latrans</u>, cf <u>Centrocercus</u> sp., <u>Erethizon</u> ? <u>dorsatum</u>, †<u>Equus</u> ?
<u>conversidens</u>, †<u>Equus</u> sp. (large), †<u>Glossotherium</u> cf <u>harlani</u>,
<u>Lepus</u> cf <u>townsendi</u>, <u>Lynx</u> cf <u>canadensis</u>, †<u>Mammuthus</u> cf <u>colum-</u>
<u>bi</u>, <u>Microtus</u> <u>montanus</u>, <u>Mustela</u> cf <u>erminea</u>, <u>M</u>. <u>vison</u>, <u>Ondatra</u>
sp., <u>Peromyscus</u> <u>maniculatus</u>, <u>Phenacomys</u> <u>intermedius</u>, †<u>Smilo-</u>
<u>don</u> cf <u>fatalis</u>, <u>Sorex</u> <u>palustris</u>, <u>Spermophilus</u> cf <u>armatus</u>,

Tamias minimus, Taxidea taxus, Thomomys talpoides

285 Slaughter Canyon, Eddy Co., NM. Kurten, 1975. 174

286 Smith Creek Cave, Reddish-brown silt zone, White Pine
Co., NV. 1950 m, 6123 m ee. 12,600 ± 170 (A-1565).
Thompson and Mead, 1982.
Ammospermophilus leucurus, Antilocapra americana, ? Antrozous
pallidus, Bassariscus astutus, ? Bison, Brachylagus idahoensis, †Camelops sp., Canis cf latrans, C. lupus, ? †Capromeryx
sp., Centrocercus urophasianus, Cervus elaphus, Dipodomys cf
ordi, †Equus sp. (large), †Equus sp. (small), Erethizon dorsatum, Felis concolor, ? †Hemiauchenia sp., Lepus sp., Lynx
rufus, Marmota flaviventris, †Martes nobilis, Microdipodops
cf megacephalus, Microtus cf longicaudus, M. cf montanus,
Mustela erminea, M. vison, Neotoma cinerea, N. lepida, Ochotona princeps, ? Odocoileus sp., †Oreamnos harringtoni, Ovis
canadensis, †Panthera ? leo atrox, P. onca, Peroganthus cf
parvus, Peromyscus sp., Phenacomys cf intermedius, Soricidae,
Spermophilus cf beldingi, S. cf lateralis, S. cf richardsoni,
Spilogale putorius, Sylvilagus sp., Tamias cf umbrinus, T.
minimus, Taxidea taxus, Thomomys cf talpoides, ? Ursus sp.,
Vulpes velox, V. vulpes

287 Smith Creek Cave No. 4, White Pine Co., NV. 1950 m, 6123
m ee. 12,235 ± 395 (GX-5863). Thompson and Mead, 1982.
Ochotona cf princeps

288 Smith Creek Cave No. 5, White Pine Co., NV. 1950 m, 6123
m ee. 13,340 ± 430 (A-2094). Thompson and Mead, 1982.
Ochotona cf princeps

289 **Smoky River,** Hwy 34 Crossing, Alberta. Churcher and
Wilson, 1979. 74

290 **South Chimney,** UTEP Loc. 7, Dry Cave, Eddy Co., NM.
Interstadial. UTEP. 217

291 **South of Charleston,** Cochise Co., AZ. Lindsay and
Tessman, 1974. 72

292 **Springville,** Utah Co., UT. Miller, 1976. 134

293 **Stalag 17,** UTEP Loc. 23, Dry Cave, Eddy Co., NM. 1280 m,
4758 m ee. 11,880 ± 250 (I-5987). Harris, 1977<u>b</u>; UTEP.
†Camelops hesternus, †Equus conversidens, †E. niobrarensis,
Lagurus curtatus, Lepus sp., Microtus cf longicaudus, M. mex-
icanus, Neotoma cinerea, N. floridana, Odocoileus ? hemionus,
Pappogeomys castanops, Pitymys ochrogaster, Sorex merriami,
Sylvilagus nuttalli, Vulpes sp.

294 **Stanton Cave,** Coconino Co., AZ. 927 m, 4779 m ee.
Series of dates from >35,000 (UA-A1056) to 2450 ± 80
(UA-A1165). Four dates from >20 cm are 10,760 ± 200
(UA-A1154) to 13,770 ± 500 (UA-A1132). Sediments above 20 cm
date <6000 BP. Euler, 1978.
Bison sp., †Camelops sp., Centrocercus urophasianus, Eptesi-
cus fuscus, Lutra canadensis, Myotis sp., Neotoma sp., Notio-
sorex crawfordi, †Oreamnos harringtoni, Ovis canadensis,
Peromyscus sp., Sylvilagus sp.

295 **Stauffer Chemical Plant,** Sweetwater Co., WY. Anderson,
1974. 27

296 Streamview No. 1, Snake Range, White Pine Co., NV. 1860 m, 6033 m ee. 11,010 ± 400 (A-2095). Thompson and Mead, 1982.
Ochotona sp.

297 Streamview No. 2, Snake Range, White Pine Co., NV. 1860 m, 6033 m ee. 17,350 ± 435 (GX-5866). Thompson and Mead, 1982.
Ochotona sp.

298 Taber, Alberta. Correlation suggests >32,000 BP. Harington, 1978. 109

299 Tank Trap Wash, El Paso Co., TX. >33,000 (A-1721). Van Devender and Riskind, 1979. 149

300 Taylor, B.C. †Mammuthus primigenius tooth dated 27,400 ± 580 (GSC-2034). Churcher and Wilson, 1979. 3, 17, 136

301 Teichart Gravel Pit, Sacramento Co., CA. Late Pleistocene. Nowak, 1979. 31, 34

302 Thatcher Basin, Caribou and Franklin counties, ID. †Mammuthus and Marmota from Main Canyon Fm.; ^{14}C date on mollusks indicates formation from >33,700 to 27,000; rabbit and ground squirrel <27,000 BP. Bright, 1967. 116, 129, 134, 137, 289

303 Tombstone Gulch, Cochise Co., AZ. Lindsay and Tessman, 1974. 133

304 Tooth Cave, Coconino Co., AZ. Ca. 1400 m. DeSaussure,

1956. 72, 138, 194

305 Touchet, Walla Walla Co., WA. Newcomb and Repenning,
1970. 133

306 Touchet Beds, Benton Co., WA. Newcomb and Repenning,
1970. 133

307 Tranquility, Fresno Co., CA. Wisconsinan. Nowak, 1979.
34

308 Tree Springs, Sandia Mts., Bernalillo Co. ?, NM. 2550 m,
6295 m ee. Martin and Guilday, 1967. 131

309 Tse'an Bida, Coconino Co., AZ. Mead, 1981. 205

310 Tse-an Kaetan, Coconino Co., AZ. Ca. 1400 m.
DeSaussure, 1956; Lange, 1956. 138, 203

311 Tse-an Olje, Coconino Co., AZ. Ca. 1400 m. Lange,
1956. 138

312 TT II, UTEP Loc. 54, Dry Cave, Eddy Co., NM. 1280 m,
4758 m ee. 10,730 ± 150 (I-6200). Harris, 1977b; UTEP.
†Bison cf antiquus, Cryptotis parva, Dipodomys spectabilis,
Eptesicus fuscus, †Equus conversidens, †Equus niobrarensis,
†Hemiauchenia sp., Lagurus curtatus, Lepus sp., Microtus
mexicanus, M. pennsylvanicus, Mustela frenata, Myotis sp.,
Neotoma cf cinerea, N. cf floridana, N. micropus, Notiosorex
crawfordi, Onychomys leucogaster, Perognathus sp., Peromyscus
sp., Pitymys ochrogaster, Sigmodon sp., Sorex merriami, S.

monticolus/vagrans, Sylvilagus nuttalli, Thomomys bottae, Vulpes velox

313 Tucson Brickyard, Pima Co., AZ. Lindsay and Tessman, 1974. 17, 72, 133

314 Tucson Mts., Pima Co., AZ. Lindsay, 1978. 17, 231, 289, 308, 323, 338

315 Tucson Mts. No. 1, Pima Co., AZ. 890 m. 12,430 ± 400 (A-1195). Lindsay and Tessman, 1974. 231, 308, 323, 338

316 Tulare Lake, Kings Co., CA. Ca. 58 m. Riddell and Olsen, 1969. 17, 72, 104

317 Tularosa Cave, Catron Co., NM. 2061 m. Howell, 1915. 138

318 Tule Springs Unit B-2, Clark Co., NV. 703 m, 4555 m ee. >40,000 BP. Mawby, 1967.
†Bison antiquus, †Camelops hesternus, †Equus sp. (large), †Equus sp. (small), Felis sp. or Lynx sp., Lepus sp., †Mammuthus columbi, †Megalonyx sp., Microtus sp., †Nothrotheriops shastensis, †Panthera leo atrox, ? Sylvilagus sp.

319 Tule Springs Unit D, Clark Co., NV. 703 m, 4555 m ee. 22,600; 31,300 BP. Mawby, 1967. 27, 134

320 Tule Springs Unit E-1, Clark Co., NV. 703 m, 4555 m ee. Between 14,000 and 11,500 BP. Mawby, 1967.
? Brachylagus idahoensis, †Camelops hesternus, Canis latrans, Dipodomys sp., Equus cf caballus, †Equus cf E. (Asinus),

Felis (Puma) sp., Lepus sp., †Mammuthus columbi, ? †Megalonyx sp., Microtus cf californicus, Odocoileus sp., Ondatra zibethicus, Sylvilagus sp., ? †Tetrameryx, Thomomys ? bottae

321 **Twin Bridges Gravel Pit**, Alberta. McDonald, 1981. 21

322 **Twin Falls**, Twin Falls Co., ID. Wisconsinan. Nowak, 1979. 34

323 **Union Pacific Mammoth Site**, SW Rawlins, Carbon Co., WY. 11,280 ± 350 (I-449) (part). Anderson, 1974. 72, 134

324 **Upper Sloth Cave**, Culberson Co., TX. 2000 m, 5424 m ee. 11,760 ± 610 (A-1519) on artiodactyl dung; 10,780 ± 140 (A-1534) on sloth dung; 11,060 ± 180 (A-1584) on sloth dung; 10,750 ± 140 (A-1583) on sloth dung (Spaulding and Martin, 1979). Logan and Black, 1979.
Antrozous pallidus, Bassariscus astutus, Cryptotis parva, Eptesicus fuscus, Felis concolor, Lasionycteris noctivagans, Lepus cf californicus, Marmota flaviventris, Microtus mexicanus, Mustela frenata, Myotis sp., M. velifer, Neotoma albigula, N. cinerea, N. mexicana, N. micropus, †Nothrotheriops shastensis, Notiosorex crawfordi, Pappogeomys castanops, Peromyscus eremicus, Peromyscus spp., Plecotus townsendi, Sorex cinereus, Spermophilus variegatus, Sylvilagus sp., Thomomys bottae

325 **Utah Cave**, UT. Miller, 1976. cf 27

326 **Vancouver Is.**, B.C Probably various ages, most >20,000 BP. Harington, 1978. ? 17, cf 72, 131, cf 134, 135, 317

327 **Ventana Cave,** Pima Co., AZ. 750 m, 4228 m ee. 11,300 $\pm$
1200 on charcoal from volcanic debris level (Haynes, 1964).
Colbert, 1950.
Bison sp., †Canis dirus, C. latrans, C. lupus, Cervidae
(large), Cynomys ludovicianus, †Equus occidentalis, Lepus
californicus, Neotoma sp., †Nothrotheriops shastensis, †Pan-
thera leo atrox, Spermophilus lateralis, †Stockoceros conklingi (poss. Antilocapra), †Tapirus sp., Taxidea taxus,
Tayassu sp. (or †Platygonus), Vulpes macrotis

328 **Vulture Cave,** Mohave Co., AZ. 645 m, 4497 m ee. Mead
and Phillips, 1981.
Antilocapra americana, Bassariscus astutus, †Camelops cf hesternus, Chiroptera, Erethizon dorsatum, Marmota cf flaviventris, Microtus sp., Neotoma lepida, N. cf stephensi, Notiosorex crawfordi, Odocoileus sp., Ovis canadensis, Perognathus
cf intermedius, Peromyscus sp., Spermophilus variegatus, Sylvilagus sp.

329 **W.A.C. Bennett Dam,** B.C. >11,600 $\pm$?1000 (I-2244).
Churcher and Wilson, 1979. 133

330 **Wakefield,** Cochise Co., AZ. 1722 m. Lindsay and
Tessman, 1974. 133

331 **Walla Walla,** Walla Walla Co., WA. McDonald, 1981. 21

332 **Walsenburg,** Huerfano Co., CO. Cockerell, 1909. 105

333 **Wapiti River,** SW Grande Prairie, Alberta. Churcher and
Wilson, 1979. 45

334 **Warm Springs No. 1**, Silver Bow Co., MT. Kurten and Anderson, 1980; Zakrzewski, in litt. 111, 157

335 **Wasden**, Bonneville Co., ID. Mostly Holocene; some >8000 BP, based on ^{14}C date (all but <u>Marmota</u> and <u>Spilogale</u> found in latter). Guilday, 1969; Kurten and Anderson, 1980; McDonald, 1981. 21, 24, cf 66, 111, 117, 149, ? 159, cf 160, 179, cf 201, cf 228, 231, 246, 293, 298, cf 314, cf 315, 326, 343, 344

336 **Watino**, Alberta. "Post-glacial." Churcher and Wilson, 1979.
†<u>Bison</u> cf <u>alaskensis</u>, †<u>Equus</u> cf <u>conversidens</u>, †<u>E</u>. cf <u>niobra-rensis</u>, cf †<u>Hemiauchenia</u> sp., <u>Ovibos</u> cf <u>moschatus</u>

337 **Watino**, Alberta. Sangamonian or earlier. Below unconformable beds with wood date of 43,500 ± 620 (GSC-1020). Churcher and Wilson, 1979. 17, 74, 81, 133, 289

338 **Wellton Hills**, Yuma Co., AZ. Ca. 160 m. Lindsay and Tessman, 1974. 63, 175, 220, 289

339 **Wenatchee**, Chelan Co., WA. Russell, 1968. 343

340 **Whipple Gravels**, Navajo Co., AZ. Lindsay and Tessman, 1974. 72

341 **White River**, Tulare Co., CA. McDonald, 1981. 21

342 **Whitewater Draw**, Cochise Co., AZ. Hester, 1960. 17, 26, 31, 72, 134

343 Wilcox Gravel, Cochise Co., AZ. Lindsay, 1978; Lindsay and Tessman, 1974. 26, 72, 133

344 Willamette Valley, near Woodburn, Marion Co., OR. Late Pleistocene. Nowak, 1979. 31

345 Williams Cave, Culberson Co., TX. 1495 m, 4949 m ee. Dates on sloth dung: 11,930 ± 170 (A-1588); 11,140 ± 320 (A-1589); 12,100 ± 210 (A-1563). Holocene material is mixed with that of the Pleistocene. Ayer, 1936.
Antilocapra americana, †Canis cf dirus, cf Cervus elaphus, Cynomys cf gunnisoni, †Equus conversidens, Erethizon dorsatum, Felis concolor, Lepus californicus, Lynx cf rufus, Neotoma albigula, †Nothrotheriops sp., Odocoileus hemionus, O. virginianus, Ovis canadensis, Pappogeomys castanops, Perognathus cf intermedius, Spermophilus variegatus, Sylvilagus auduboni, Urocyon cinereoargenteus, cf Ursus arctos

346 Willow Creek Canyon, Humboldt Range, NV. 214

347 Wilson Butte Cave, Jerome Co., ID. Ca. 1311 m, 5859 m ee. 15,000 ± 800, 14,500 ± 500 (M-1409). Frison et al., 1978; Gruhn and Bryan, 1981; Guilday, 1969.
Bison bison, Brachylagus idahoensis, †Camelops sp., Centrocercus urophasianus, Clethrionomys gapperi, †Equus sp., cf †Hemiauchenia, Homo sapiens, Lagurus curtatus, Lepus californicus, L. townsendi, Marmota flaviventris, †Martes nobilis, Microtus longicaudus, M. montanus, Mustela erminea, M. frenata, M. nigripes, Neotoma cinerea, ? †Nothrotheriops sp., Ochotona princeps, Peromyscus ? truei, ? Phenacomys intermedius, Spermophilus armatus, S. ? beldingi, S. richardsoni,

S. townsendi, Spilogale putorius, Sylvilagus nuttalli, Thomomys talpoides

348 Wolcott No. 2, Pima Co., AZ.   Lindsay and Tessman, 1974.
63, 89, 175, 200, 220, 231, 288, 308, 338

349 Wolcott No. 4, Pima Co., AZ.   Lindsay and Tessman, 1974.
63, 220, 231

350 Wolcott No. 5, Pima Co., AZ.   Lindsay and Tessman, 1974.
63, 190, 200, 220, 231, 308, 338

351 Yarmany Station, CO.   McDonald, 1981.   21

352 Zuma Creek, Los Angeles Co., CA.   Kurten and Anderson,
1980.   332

ADDENDA

Various records have appeared or been drawn to my attention since completion of the manuscript. A few of the more pertinent are listed below with a reference or source.

Marmes Rockshelter, Snake River, Franklin Co., WA. Ca. 10,000 BP. Lyman, R. L., and S. D. Livingston. 1983. Late Quaternary mammalian zoogeography of eastern Washington. Quat. Res., 20:360-373.
Alopex lagopus, Martes americana

Jeppson Locality, Benton Co. ?, WA. Ca. 13,000 BP. Lyman and Livingston, op. cit.
Onychomys leucogaster, Perognathus sp., Spermophilus townsendi, Thomomys sp.

Lake Bonneville, shoreline deposits, Salt Lake Co., UT. ^{14}C-date of 12,650 ± 70 BP. Nelson, M. E., and J. H. Madsen, Jr. 1983. A giant short-faced bear (Arctodus simus) from the Pleistocene of northern Utah. Trans. Kansas Acad. Sci., 86: 1-9.
†Arctodus simus

Sheep Camp Shelter, San Juan Co., NM. Gillespie, W. B. In press. [Abstr] Late Quaternary small vertebrates from Chaco Canyon, northwestern New Mexico. J. New Mexico Acad. Sci.

Brachylagus idahoensis, Centrocercus urophasianus, Lagurus curtatus, Phenacomys intermedius

Umatilla Mammoth Site, Benton Co., WA. Probably shortly after 13,000 BP. Lyman and Livingston, op. cit.
Antilocapra americana, †Bootherium sp.

U-Bar Cave, Hidalgo Co., NM. Ca. 1545 m, 4916 m ee. Mixed Holocene and Pleistocene. Lambert, M. F., and J. R. Ambler. 1961. A survey and excavation of caves in Hidalgo County, New Mexico. Sch. Amer. Res. Monogr., 25:1-107; UTEP (tempo-rary repository for Museum of New Mexico specimens). Part-ial, tentative list.
†Arctodus simus, †Desmodus stocki, †Equus sp. (larger), †Equus sp. (smaller), ? †Euceratherium sp., Lagurus curtatus, †Leporidae, Marmota flaviventris, Microtus sp., Neotoma cine-rea, N. stephensi, †Nothrotheriops shastensis, Sorex merriami

LITERATURE CITED

Akersten, W. A., R. L. Reynolds, and A. E. Tejada-Flores.
1979. New mammalian records from the Late Pleistocene of
Rancho La Brea. Bull. So. California Acad. Sci., 78:141-
143.

Allen, J. A. 1913. Ontogenetic and other variations in
muskoxen, with a systematic review of the muskox group,
Recent and extinct. Mem. Amer. Mus. Nat. Hist., n.s.,
1:101-226, pls. 11-28.

Allison, I. S. 1966. Fossil Lake, Oregon, its geology and
fossil faunas. Oregon State Monogr., Stud. Geol., 9:1-48.

───── and H. A. Boyd. 1954. A fossil camel from Oregon.
Proc. Oregon Acad. Sci., 3:31.

Anderson, E. 1968. Fauna of the Little Box Elder Cave,
Converse county, Wyoming. Univ. Colorado Stud., Ser.
Earth Sci., 6:1-59.

─────. 1974. A survey of the Late Pleistocene and Holocene
mammal fauna of Wyoming. Pp. 78-87, in Applied geology
and archaeology: The Holocene history of Wyoming (M.
Wilson, ed.). Geol. Surv. Wyoming, Rep. Inv., 10.

───── and J. A. White. 1975. Caribou (Mammalia, Cervidae)
in the Wisconsinan of southern Idaho. Tebiwa, 17:59-65.

AOU. 1957. Check-list of North American birds. Amer.
Ornithol. Union, Lord Baltimore Press, Baltimore, 691 pp.

Applegarth, J. S. 1979. Herpetofauna (anurans and lizards)
of Eddy County, New Mexico: Quaternary changes and envi-

ronmental implications. Ph.D. Dissert., Univ. New Mexico, Albuquerque, 258 pp.

Arata, A. A., and J. H. Hutchison. 1964. The raccoon (Procyon) in the Pleistocene of North America. Tulane Stud. Geol., 2:21-27.

Armstrong, D. M. 1972. Distribution of mammals in Colorado. Mus. Nat. Hist., Univ. Kansas Monogr., 3:1-415.

Ayer, M. Y. 1936. The archaeological and faunal material from Williams Cave, Guadalupe Mountains, Texas. Proc. Acad. Nat. Sci. Philadelphia, 88:599-618, pl. 15.

Betancourt, J. L., and T. R. Van Devender. 1981. Holocene vegetation in Chaco Canyon, New Mexico. Science, 214:656-658.

Brackenridge, G. R. 1978. Evidence for a cold, dry full-glacial climate in the American Southwest. Quat. Res., 9:22-40.

Brattstrom, B. H. 1958. New records of Cenozoic amphibians and reptiles from California. Bull. So. California Acad. Sci., 57:5-13.

Bright, R. C. 1967. Late-Pleistocene stratigraphy in Thatcher Basin, southeastern Idaho. Tebiwa, 10:1-7.

Broecker, W. S., and A. Kaufman. 1965. Radiocarbon chronology of Lake Lahontan and Lake Bonneville, 2, Great Basin. Bull. Geol. Soc. Amer., 76:537.

_____ and J. van Donk. 1970. Insolation changes, ice volumes, and the O^{18} record in deep-sea cores. Rev. Geophysic Space Physics, 8:169-197.

Brown, D. E., and C. H. Lowe. 1980. Biotic communities of the Southwest. Rocky Mtn. For. Range Exp. Sta.: General Tech. Rep. RM-78 [map].

Bryan, K. 1950. Geologic interpretation of the deposits. Pp. 75-126, in The stratigraphy and archaeology of Ventana

Cave, Arizona (E. W. Haury and collaborators). Univ. Arizona Press and Univ. New Mexico Press, Tucson and Albuquerque, 599 pp.

Buwalda, J. P. 1914. Pleistocene beds at Manix in the eastern Mohave Desert region. Univ. California Publ., Bull. Dept. Geol., 7:443-464.

Cary, M. 1917. Life zone investigations in Wyoming. N. Amer. Fauna, 42:1-95.

Choate, J. R., and E. R. Hall. 1967. Two new species of bats, genus Myotis, from a Pleistocene deposit in Texas. Amer. Midl. Nat., 78:531-534.

Chomko, S. A. 1978. Paleoenvironmental studies at Prospects Shelter, Wyoming. AMQUA Abstr., 1978:193.

Churcher, C. S. 1968. Pleistocene ungulates from the Bow River gravels at Cochrane, Alberta. Canadian J. Earth Sci., 5:1467-1488.

————. 1972. Imperial mammoth and Mexican half-ass from near Bindloss, Alberta. Canadian J. Earth Sci., 9:1562-1567.

———— and M. Wilson. 1979. Quaternary mammals from the eastern Peace River district, Alberta. J. Paleontol., 53:71-76.

Cockerell, T. D. A. 1909. A fossil ground-sloth in Colorado. Univ. Colorado Stud., 6:309-312, pls. 1-2.

Colbert, E. H. 1950. The fossil vertebrates. Pp. 126-148, in The stratigraphy and archaeology of Ventana Cave, Arizona (E. W. Haury and collaborators). Univ. Arizona Press and Univ. New Mexico Press, Tucson and Albuquerque, 599 pp.

Conkling, R. P. 1932. Conkling Cavern: The discoveries in the bone cave at Bishops Cap, New Mexico. West Texas Hist. Soc. Bull., 44:39-41.

Cosgrove, C. B. 1947. Caves of the Upper Gila and Hueco areas in New Mexico and Texas. Papers Peabody Mus. Amer. Archeol. Ethnol., Harvard Univ., 24(2):1-182.

Cushing, E. J. 1965. Problems in the Quaternary phytogeography of the Great Lakes region. Pp. 403-416, in The Quaternary of the United States (H. E. Wright, Jr., and D. G. Frey, eds.). Princeton Univ. Press, Princeton, 922 pp.

DeSaussure, R. 1956. Remains of California condor in Arizona caves. Plateau, 29:44-45.

Diersing, V. E. 1979. Noteworthy records of Merriam's shrew, Sorex merriami, from New Mexico. Southwestern Nat., 24:708-709.

Dorsey, S. L. 1977. A reevaluation of two new species of fossil bats from Inner Space Caverns. Texas J. Sci., 28: 103-108.

Downs, T., H. Howard, T. Clements, and G. A. Smith. 1959. Quaternary animals from Schuiling Cave in the Mojave Desert, California. Los Angeles Co. Mus., Contrib. Sci., 29:1-21.

Durrant, S. D., and N. K. Dean. 1961. Mammals of Navajo Reservoir Basin in Colorado and New Mexico, 1960. Pp. 155-182, in Ecological studies of the flora and fauna of Navajo Reservoir Basin, Colorado and New Mexico. Univ. Utah, Anthrop. Papers, 55:1-203.

Elftman, H. O. 1931. Pleistocene mammals of Fossil Lake, Oregon. Amer. Mus. Nov., 481:1-21.

Emiliani, C., and N. J. Shackleton. 1974. The Brunhes Epoch: Isotopic paleotemperatures and geochronology. Science, 183:511-514.

Euler, R. C. 1978. Archaeological and paleobiological studies at Stanton's Cave, Grand Canyon National Park, Arizona--A report of progress. Natl. Geogr. Soc. Res. Rep.,

1969 Proj.:141-162.

Ferdon, E. N., Jr. 1946. An excavation of Hermit's Cave, New Mexico. School Amer. Res., Monogr., 10:1-29.

Findley, J. S. 1965. Shrews from Hermit Cave, Guadalupe Mountains, New Mexico. J. Mamm., 46:206-210.

————, A. H. Harris, D. E. Wilson, and C. Jones. 1975. Mammals of New Mexico. Univ. New Mexico Press, Albuquerque, 360 pp.

Foreman, F., K. H. Clisby, and P. B. Sears, with a discussion by C. E. Stearns. 1959. Plio-Pleistocene sediments and climates of the San Augustine Plains, New Mexico. New Mexico Geol. Soc., 10th Field Conf.:117-120.

Frison, G. C., and D. N. Walker. 1978. The archeology of Little Canyon Creek Cave and its associated Late Pleistocene fauna. AMQUA Abstr., 1978:200.

————, D. N. Walker, S. D. Webb, and G. M. Zeimens. 1978. Paleo-Indian procurement of Camelops on the northwestern plains. Quat. Res., 10:385-400.

Geist, V. 1971. Mountain sheep: A study in behavior and evolution. Univ. Chicago Press, Chicago, 383 pp.

Ginn, H. 1973. Little, pygmy and elf owls. Pp. 164-185, in Owls of the world, their evolution, structure and ecology (J. A. Burton, ed.). E. P. Dutton, New York, 216 pp.

Graham, R. 1959. Additions to the Pleistocene fauna of Samwel Cave, California. I. Canis lupus and Canis latrans. Cave Stud., 10:54-67.

Graham, R. W. 1981. Preliminary report on Late Pleistocene vertebrates from the Selby and Dutton archaeological/paleontological sites, Yuma County, Colorado. Contrib. Geol., Univ. Wyoming , 20:33-56.

Gruhn, R. 1961. The archaeology of Wilson Butte Cave, south-central Idaho. Occas. Papers Idaho State College

Mus., 6:1-198.

———— and A. L. Bryan. 1981. A response to McGuire's cautionary tale about the association of man and extinct fauna in Great Basin cave sites. Quat. Res., 16:117-121.

Guilday, J. E. 1967. Notes on the Pleistocene big brown bat Eptesicus fuscus (Brown). Ann. Carnegie Mus., 39:105-114.

————. 1969. Small mammal remains from the Wasden Site (Owl Cave), Bonneville County, Idaho. Tebiwa, 12:47-57.

———— and E. K. Adam. 1967. Small mammal remains from Jaguar Cave, Lemhi County, Idaho. Tebiwa, 10:26-36.

————, H. W. Hamilton, and E. K. Adam. 1967. Animal remains from Horned Owl Cave, Albany County, Wyoming. Univ. Wyoming Contrib. Geol., 6:97-99.

Guthrie, R. D. 1968. Paleoecology of the large-mammal community in interior Alaska during the Late Pleistocene. Amer. Midland Nat., 79:346-363.

Hager, M. W. 1972. A late Wisconsin-Recent vertebrate fauna from the Chimney Rock Animal Trap, Larimer County, Colorado. Univ. Wyoming Contrib. Geol., 11:63-71.

————. 1974. Late Pliocene and Pleistocene history of the Donnelly Ranch vertebrate site, southeastern Colorado. Univ. Wyoming Contrib. Geol., Spec. Paper, 2:1-62.

Hall, E. R. 1960. Small carnivores from San Josecito Cave (Pleistocene), Nuevo Leon, Mexico. Univ. Kansas Publ., Mus. Nat. Hist., 9:531-538.

————. 1981. The mammals of North America. John Wiley & Sons, New York, 2 vol.

———— and K. R. Kelson. 1959. The mammals of North America. Ronald Press, New York, 2 vol.

Harington, C. R. 1968. A Pleistocene muskox (Symbos) from Dease Lake, British Columbia. Canadian J. Earth Sci., 5: 1161-1165 + 3 pls.

————. 1969. Pleistocene remains of the lion-like cat (Panthera atrox) from the Yukon Territory and northern Alaska. Canadian J. Earth Sci., 6:1277-1288.

————. 1972. Extinct animals of Rampart Cave. Canadian Geogr. J., 85:178-183.

————. 1978. Quaternary vertebrate faunas of Canada and Alaska and their suggest chronological sequence. Syllogeus, 15:1-105.

Harrington, M. R. 1933. Gypsum Cave, Nevada. Southwest Mus. Papers, 8:1-197.

Harris, A. H. 1963a. Ecological distribution of some vertebrates in the San Juan Basin, New Mexico. Mus. New Mexico, Papers Anthrop., 8:1-64.

————. 1963b. Vertebrate remains and past environmental reconstruction in the Navajo Reservoir District. Mus. New Mexico Papers Anthrop., 11:1-71.

————. 1970a. Past climate of the Navajo Reservoir District. Amer. Antiq., 35:374-377.

————. 1970b. The Dry Cave mammalian fauna and late pluvial conditions in southeastern New Mexico. Texas J. Sci., 22:3-27.

————. 1977a. Appendix A. Faunal remains from Chimney Rock Mesa. Pp. 73-76, in Archaeological investigations at Chimney Rock Mesa: 1970-1972 (F. W. Eddy, ed.). Mem. Colorado Archaeol. Soc., 1:1-91.

————. 1977b. Wisconsin age environments in the northern Chihuahuan Desert: Evidence from the higher vertebrates. Pp. 23-52, in Transactions of the symposium on the biological resources of the Chihuahuan Desert Region, United States and Mexico (R. H. Wauer and D. H. Riskind, eds.). Natl. Park Serv., Trans. Proc. Ser., 3:1-658.

————. 1981. [Review of] Pleistocene mammals of North

America. J. Mamm., 62:653-654.

_____. In press, a. Neotoma in the Late Pleistocene of New Mexico and Chihuahua. In, Carnegie Mus. Nat. Hist., Spec. Publ., 8.

_____. In press, b. Two new species of Late Pleistocene woodrats (Cricetidae:Neotoma) from New Mexico. J. Mamm.

_____ and C. R. Crews. 1983. Conkling's roadrunner--a subspecies of the California roadrunner? Southwestern Nat., 28:407-412.

_____ and J. S. Findley. 1964. Pleistocene-Recent fauna of the Isleta Caves, Bernalillo County, New Mexico. Amer. J. Sci., 262:114-120.

_____ and P. Mundel. 1974. Size reduction in bighorn sheep (Ovis canadensis) at the close of the Pleistocene. J. Mamm., 55:678-680.

_____ and L. S. W. Porter. 1980. Late Pleistocene horses of Dry Cave, Eddy County, New Mexico. J. Mamm., 61:46-65.

_____, R. A. Smartt, and W. R. Smartt. 1973. Cryptotis parva from the Pleistocene of New Mexico. J. Mamm., 54: 512-513.

Haury, E. W. 1950. The stratigraphy and archaeology of Ventana Cave, Arizona. Univ. Arizona Press and Univ. New Mexico Press, Tucson and Albuquerque, 599 pp.

Hay, O. P., and H. J. Cook. 1930. Fossil vertebrates collected near, or in association with, human artifacts at localities near Colorado, Texas; Frederick, Oklahoma; and Folsom, New Mexico. Colorado Mus. Nat. Hist., Proc., 9: 4-40, pls. 1-14.

Haynes, C. V., Jr. 1964. Fluted projectile points: Their age and dispersion. Science, 145:1408-1413.

_____. 1967. Quaternary geology of the Tule Springs area, Clark County, Nevada. Pp. 16-104, in Pleistocene studies in southern Nevada (M. Wormington and Ellis, eds.). Nevada State Mus., Anthrop. Papers, 13:1-411.

————. 1968. Preliminary report on the late Quaternary geology of the San Pedro Valley, Arizona. Arizona Geol. Soc., Southern Arizona Guidebook, 3:79-96.

————. 1975. Pleistocene and Recent stratigraphy. Pp. 57-96, in Late Pleistocene environments of the Southern High Plains (F. Wendorf and J. J. Hester, eds.). Publ. Fort Burgwin Res. Cent., 9:1-290.

———— and G. A. Agogino. 1966. Prehistoric springs and geochronology of the Clovis Site, New Mexico. Amer. Antiq., 31:812-821.

Hennings, D., and R. S. Hoffman. 1977. A review of the taxonomy of the Sorex vagrans species complex from western North America. Occas. Papers Mus. Nat. Hist., Univ. Kansas, 68:1-35.

Hester, J. J. 1960. Late Pleistocene extinction and radiocarbon dating. Amer. Antiq., 26:58-77.

————. 1967. The agency of man in animal extinctions. Pp. 169-192, in Pleistocene extinctions, the search for a cause (P. S. Martin and H. E. Wright, Jr., eds.). Yale Univ. Press, New Haven, 453 pp.

Heusser, C. J. 1965. A Pleistocene phytogeographical sketch of the Pacific Northwest and Alaska. Pp. 469-483, in The Quaternary of the United States (H. E. Wright, Jr., and D. G. Frey, eds.). Princeton Univ. Press, Princeton, 922 pp.

Hibbard, C. W., and B. A. Wright. 1956. A new Pleistocene bighorn sheep from Arizona. J. Mamm., 37:105-107.

————, C. E. Ray, D. E. Savage, D. W. Taylor, and J. E. Guilday. 1965. Quaternary mammals of North America. Pp. 509-525, in The Quaternary of the United States (H. E. Wright, Jr., and D. G. Frey, eds.). Princeton Univ.

Press, Princeton, 922 pp.

Hibben, F. C. 1941. Evidences of early occupation in Sandia Cave, New Mexico, and other sites in the Sandia-Manzano Region. Smithsonian Misc. Coll., 99(23):1-65.

Hoffmeister, D. F., and L. de la Torre. 1960. A revision of the wood rat Neotoma stephensi. J. Mamm., 41:476-491.

Holman, J. A. 1970. A Pleistocene herpetofauna from Eddy County, New Mexico. Texas J. Sci., 22:29-39.

Hopkins, M. L., R. Bonnichsen, and D. Fortsch. 1969. The stratigraphic position and faunal associates of Bison (Gigantobison) latifrons in southeastern Idaho, a progress report. Tebiwa, 12:1-7.

Howard, E. B. 1932. Caves along the slopes of the Guadalupe Mountains. Texas Archeol. Paleontol. Soc., Bull., 4:7-20.

Howard, H. 1971. Quaternary avian remains from Dark Canyon Cave, New Mexico. Condor, 73:237-240.

Howell, A. H. 1915. Revision of the American marmots. N. Amer. Fauna, 37:1-80.

Hutchison, J. H. 1967. A Pleistocene vampire bat (Desmodus stocki) from Potter Creek Cave, Shasta County, California. Paleobios, 3:1-6.

Jacobson, G. L., Jr., and R. H. W. Bradshaw. 1981. The selection of sites for paleovegetational studies. Quat. Res., 16:80-96.

Jakway, G. E. 1958. Pleistocene Lagomorpha and Rodentia from the San Josecito Cave, Nuevo Leon, Mexico. Trans. Kansas Acad. Sci., 61:313-327.

Jelinek, A. J. 1957. Pleistocene faunas and early man. Papers Michigan Acad. Sci., Arts, Lett., 42:225-237.

Johnsgard, P. A. 1973. Grouse and quails of North America. Univ. Nebraska Press, Lincoln, 553 pp.

Jones, C. J. 1961. Additional records of shrews in New Mexico. J. Mamm., 42:399.

Jones, J. K., Jr. 1958. Pleistocene bats from San Josecito Cave, Nuevo Leon, Mexico. Univ. Kansas Publ., Mus. Nat. Hist., 9:389-396.

———. 1964. Distribution and taxonomy of mammals of Nebraska. Univ. Kansas Publ., Mus. Nat. Hist., 16:1-356.

——— et al. 1982. Revised checklist of North American mammals north of Mexico, 1982. Occas. Papers, The Museum, Texas Tech Univ., 80:1-22.

Kellogg, L. 1912. Pleistocene rodents of California. Univ. California Publ., Bull. Dept. Geol., 7:151-168.

Kuchler, A. W. 1970. Potential natural vegetation. Pp. 89-92, in The national atlas of the United States of America. U.S. Dept. Int., Geol. Surv., Washington, D.C., 417 pp.

Kurten, B. 1973. Pleistocene jaguars in North America. Comment. Biol., 62:1-23.

———. 1975. A new Pleistocene genus of American mountain deer. J. Mamm., 56:507-508.

——— and E. Anderson. 1980. Pleistocene mammals of North America. Columbia Univ. Press, New York, 442 pp.

Lance, J. F. 1953. Artifacts with mammoth remains, Naco, Arizona. III. Description of the Naco mammoth. Amer. Antiq., 19:19-22.

Lang, R., and A. H. Harris. In press. The faunal remains from Arroyo Hondo Pueblo, New Mexico--A study in short term subsistence changes. Sch. Amer. Res. Press, Arroyo Hondo Ser.

Lange, A. L. 1956. Woodchuck remains in northern Arizona caves. J. Mamm., 37:289-291.

Laudermilk, J. D., and P. A. Munz. 1934. Plants in the dung of Nothrotherium from Gypsum Cave, Nevada. Carnegie Inst. Washington Publ. 453, Paper, 4:29-37.

Lawrence, B. 1960. Fossil _Tadarida_ from New Mexico. J.
   Mamm., 41:320-322.

Leopold, A. S. 1959. Wildlife of Mexico. Univ. California
   Press, Berkeley, 568 pp.

Leopold, E. B., R. Nickmann, J. E. Hedges, and J. R. Ertel.
   1982. Pollen and lignin records of late Quaternary vegetation, Lake Washington. Science, 218:1305-1307.

Lindeborg, R. G. 1960. The desert shrew, Notiosorex, in San
   Miguel County, New Mexico. Southwestern Nat., 5:108-109.

Lindsay, E. H. 1978. Late Cenozoic vertebrate faunas,
   southeastern Arizona. New Mexico Geol. Soc. Guidebook,
   29th Field Conf., Land of Cochise, 1978:269-275.

_____ and N. T. Tessman. 1974. Cenozoic vertebrate localities and faunas in Arizona. J. Arizona Acad. Sci., 9:
   3-24.

Logan, L. E. 1981. The mammalian fossils of Muskox Cave,
   Eddy County, New Mexico. Proc. Eighth Internat. Congr.
   Speol., 1:159-160.

_____. 1983. Paleoecological implications of the mammalian
   fauna of Lower Sloth Cave, Guadalupe Mountains, Texas.
   Natl. Speol. Soc. Bull., 45:3-11.

_____ and C. C. Black. 1979. The Quaternary vertebrate
   fauna of Upper Sloth Cave, Guadalupe Mountains National
   Park, Texas. Pp. 141-158, _in_ Biological investigations in
   the Guadalupe Mountains National Park, Texas (H. H. Genoways and R. J. Baker, eds.). Natl. Park Serv., Proc.
   Trans. Ser., 4:1-442.

Long, A., and P. S. Martin. 1974. Death of American ground
   sloths. Science, 186:638-640.

Long, C. A. 1965. The mammals of Wyoming. Univ. Kansas
   Publ., Mus. Nat. Hist., 14:493-758.

_____. 1971. Significance of the Late Pleistocene fauna

from Little Box Elder Cave, Wyoming, to studies of zoogeography of Recent mammals. Great Basin Nat., 31:93-105.

Lowe, C. H., ed. 1964. The vertebrates of Arizona. Univ. Arizona Press, Tucson, 132 pp.

Lull, R. S. 1929. A remarkable ground sloth. Yale Univ., Peabody Mus. Mem., 3, pt. 2:1-39.

Lundelius, E. L., Jr. 1972a. Fossil vertebrates from the Late Pleistocene Ingleside fauna, San Patricio County, Texas. Bur. Econ. Geol., Rep. Invest., 77:1-74.

————. 1972b. Vertebrate remains from the gray sand. Pp. 148-163, in Blackwater Locality No. 1. A stratified early man site in eastern New Mexico. Fort Burgwin Res. Cent. Publ., 8:1-238.

————. 1979. Post-Pleistocene mammals from Pratt Cave and their environmental significance. Pp. 239-258, in Biological investigations in the Guadalupe Mountains National Park, Texas (H. H. Genoways and R. J. Baker, eds.). Natl. Park Serv., Proc. Trans. Ser., 4:1-422.

Lyon, G. M. 1938. Megalonyx milleri, a new Pleistocene ground sloth from southern California. Trans. San Diego Soc. Nat. Hist., 9:15-30.

MacDonald, J. R. 1967. The Maricopa brea. Mus. Alliance Quart., Los Angeles Co. Mus. Nat. Hist., 6:21-24.

McDonald, J. G., and E. Anderson. 1975. A Late Pleistocene vertebrate fauna from southeastern Idaho. Tebiwa, 18: 19-37.

McDonald, J. N. 1981. North American bison. Their classification and evolution. Univ. California Press, Berkeley, 316 pp.

McGuire, K. R. 1980. Cave sites, faunal analysis, and big-game hunters of the Great Basin: A caution. Quat. Res., 14:263-268.

Magish, D. P., and A. H. Harris. 1976. Fossil ravens from the Pleistocene of Dry Cave, Eddy County, New Mexico. Condor, 78:399-404.

Maher, L. J. 1961. Pollen analysis and postglacial vegetation history in the Animas Valley Region, southern San Juan Mountains, Colorado. Ph.D. Thesis, Univ. Minnesota, 85 pp.

––––––. 1963. Pollen analyses of surface materials from the southern San Juan Mountains, Colorado. Geol. Soc. Amer. Bull., 74:1485-1504.

Martin, L. D., and B. M. Gilbert. 1978. Excavations at Natural Trap Cave. Trans. Nebraska Acad. Sci., 6:107-116.

––––––, G. M. Gilbert, and D. B. Adams. 1977. A cheetah-like cat in the North American Pleistocene. Science, 195: 981-982.

Martin, P. S., and J. E. Guilday. 1967. A bestiary for Pleistocene biologists. Pp. 1-62, in Pleistocene extinctions, the search for a cause (P. S. Martin and H. E. Wright, Jr., eds.). Yale Univ. Press, New Haven, 453 pp.

–––––– and P. J. Mehringer, Jr. 1965. Pleistocene pollen analysis and biogeography of the Southwest. Pp. 443-451, in The Quaternary of the United States (H. E. Wright, Jr., and D. G. Frey, eds.). Princeton Univ. Press, Princeton, 992 pp.

Mawby, J. E. 1967. Fossil vertebrates of the Tule Springs Site, Nevada. Pp. 105-128, in Pleistocene studies in southern Nevada (M. Wormington and Ellis, eds.). Nevada State Mus., Anthrop. Papers, 13:1-411.

Mead, J. I. 1981. The last 30,000 years of faunal history within the Grand Canyon, Arizona. Quat. Res., 15:311-326.

–––––– and A. M. Phillips, III. 1981. The Late Pleistocene and Holocene fauna and flora of Vulture Cave, Grand Canyon, Arizona. Southwestern Nat., 26:257-288.

―――, C. V. Haynes, and B. B. Huckell. 1979. A Late Pleistocene mastodon (Mammut americanum) from the Lehner Site, southeastern Arizona. Southwestern Nat., 24:231-238.

―――, R. S. Thompson, and T. R. Van Devender. 1982. Late Wisconsinan and Holocene fauna from Smith Creek Canyon, Snake Range, Nevada. Trans. San Diego Soc. Nat. Hist., 20:1-26.

Mears, B., Jr. 1981. Periglacial wedges and the Late Pleistocene environment of Wyoming's intermontane basins. Quat. Res., 15:171-198.

Mehringer, P. J., Jr. 1965. Late Pleistocene vegetation in the Mohave Desert of southern Nevada. J. Arizona Acad. Sci., 3:172-188.

―――. 1967. The environment of extinction of the Late-Pleistocene megafauna in the arid southwestern United States. Pp. 247-266, in Pleistocene extinctions, the search for a cause (P. S. Martin and H. E. Wright, Jr., eds.). Yale Univ. Press, New Haven, 453 pp.

――― and C. W. Ferguson. 1969. Pluvial occurrence of bristlecone pine (Pinus aristata) in a Mohave Desert mountain range. J. Arizona Acad. Sci., 5:284-292.

Merriam, J. C. 1915. An occurrence of mammalian remains in a Pleistocene lake deposit at Astor Pass, near Pyramid Lake, Nevada. Univ. California Publ., Bull. Dept. Geol., 8:377-384.

Merritt, J. F. 1978. Peromyscus californicus. Mamm. Sp., 85:1-6.

Metcalf, A. L. 1970. Late Pleistocene (Woodfordian) gastropods from Dry Cave, Eddy County, New Mexico. Texas J. Sci., 22:41-46.

Miller, W. E. 1968. Occurrence of a giant bison, Bison latifrons, and a slender-limbed camel, Tanupolama, at Rancho La Brea. Contrib. Sci., Los Angeles Co. Mus. Nat. Hist., 147:1-9.

_____. 1971. Pleistocene vertebrates of the Los Angeles Basin and vicinity (exclusive of Rancho La Brea). Los Angeles Co. Mus. Nat. Hist., Sci. Bull., 10:1-124.

_____. 1976. Late Pleistocene vertebrates of the Silver Creek local fauna from north central Utah. Great Basin Nat., 36:387-424.

Mohlhenrich, J. S. 1961. Distribution and ecology of the hispid and least cotton rats in New Mexico. J. Mamm., 42:13-24.

Moodie, K. B., and T. R. Van Devender. 1979. Extinction and extirpation in the herpetofauna of the Southern High Plains with emphasis on Geochelone wilsoni (Testudinidae). Herpetologica, 35:198-206.

Nelson, E. W. 1909. The rabbits of North America. N. Amer. Fauna, 29:1-314.

Newcomb, R. C., and C. A. Repenning. 1970. Occurrence of mammoth fossils in the Touchet beds, south-central Washington. Northwest Sci., 44:16-18.

NOAA. 1974. Climatic atlas of the United States. U.S. Dept. Commerce, 80 pp.

Nowak, R. M. 1979. North American Quaternary Canis. Univ. Kansas, Monogr. Mus. Nat. Hist., 6:1-154.

O'Gara, B. W. 1978. Antilocapra americana. Mamm. Sp., 90:1-7.

Oldfield, F., and J. Schoenwetter. 1975. Discussion of the pollen analytical evidence. Pp. 149-177, in Late Pleistocene environments of the Southern High Plains (F. Wendorf and J. J. Hester, eds.). Publ. Fort Burgwin Res. Cent.,

9:1-290.

Orr, P. C. 1956. Pleistocene man in Fishbone Cave, Pershing County, Nevada. Nevada State Mus., Bull., 2:1-20.

————. 1969. Felis trumani. A new radiocarbon dated cat skull from Crypt Cave, Nevada. Santa Barbara Mus. Nat. Hist., Dept. Geol. Bull., 2:1-8.

————. 1972. The eighth Lake Lahontan (Nevada) expedition, 1957. Natl. Geogr. Soc. Res. Rep., 1955-1960 Proj.:123-126.

Porter, L. S. W. 1978. Pleistocene pluvial climates as indicated by present day climatic parameters of Cryptotis parva and Microtus mexicanus. J. Mamm., 59:330-338.

Ray, C. E., and D. E. Wilson. 1979. Evidence for Macrotus californicus from Terlingua, Texas. Occas. Papers Mus., Texas Tech Univ., 57:1-10.

Rea, A. M. 1980. Late Pleistocene and Holocene turkeys in the Southwest. Contrib. Sci., Nat. Hist. Mus. Los Angeles Co., 330:209-224.

Repenning, C. A. 1983. Pitymys meadensis Hibbard from the Valley of Mexico and the classification of North American species of Pitymys (Rodentia: Cricetidae). J. Vert. Paleontol., 2:471-482.

Riddell, F. A., and W. H. Olsen. 1969. An early man site in the San Joaquin Valley, California. Amer. Antiq., 34:121-130.

Roberts, M. F. 1970. Late glacial and postglacial environments in southeastern Wyoming. Paleogeogr., Paleoclim., Paleoecol., 7:5-19.

Russell, I. C. 1885. Geological history of Lake Lahontan, a Quaternary lake of northwestern Nevada. U.S. Geol. Surv. Monogr., 11:1-288.

Russell, R. J. 1968. Evolution and classification of the

pocket gophers of the subfamily Geomyinae. Univ. Kansas Publ., Mus. Nat. Hist., 16:473-579.

Savage, D. E. 1951. Late Cenozoic vertebrates of the San Francisco Bay Region. Univ. California Publ., Bull. Dept. Geol. Sci., 28:215-314.

Schultz, C. B., and E. B. Howard. 1935. The fauna of Burnet Cave, Guadalupe Mountains, New Mexico. Proc. Acad. Nat. Sci. Philadelphia, 87:273-298.

————, L. D. Martin, and L. G. Tanner. 1970. Mammalian distribution in the Great Plains and adjacent areas from 14,000 to 9,000 years ago. AMQUA Abstr., 1st Meeting, 1970:119-120.

Schultz, J. R. 1938. A late Quaternary mammal fauna from the tar seeps of McKittrick, California. Carnegie Inst. Washington Publ., 487:111-215.

Shimer, J. A. 1972. Field guide to landforms in the United States. Macmillan Co., New York, 272 pp.

Simons, E. L., and H. L. Alexander. 1964. Age of the Shasta ground sloth from Aden Crater, New Mexico. Amer. Antiq., 29:390-391.

Simpson, G. G. 1963. A new record of *Euceratherium* or *Preptoceras* (extinct Bovidae) in New Mexico. J. Mamm., 44:583-584.

Sinclair, W. J. 1904. The exploration of Potter Creek Cave. Univ. California Publ., Amer. Archaeol. Ethnol., 2:1-27.

Skinner, M. F. 1942. The fauna of Papago Springs Cave, Arizona, and a study of *Stockoceros*; with three new antilocaprines from Nebraska and Arizona. Bull. Amer. Mus. Nat. Hist., 80:143-220.

Slaughter, B. H. 1961. The significance of Dasypus bellus (Simpson) in Pleistocene local faunas. Texas J. Sci., 13:311-315.

————. 1975. Ecological interpretation of the Brown Sand Wedge local fauna. Pp. 179-192, in Late Pleistocene environments of the Southern High Plains (F. Wendorf and J. J. Hester, eds.). Publ. Fort Burgwin Res. Cent., 9: 1-290.

Smartt, R. A. 1977. The ecology of Late Pleistocene and Recent Microtus from south-central and southwestern New Mexico. Southwestern Nat., 22:1-19.

————— and A. H. Harris. 1979. A record of spruce (Pinaceae: Picea) from the Pleistocene of south-central New Mexico. Southwestern Nat., 24:710.

Smith, V. J. 1934. Hord Rock Shelter. Texas Archeol. Paleontol. Soc., 6:97-106.

Spaulding, W. G. 1977. Late Quaternary vegetational change in the Sheep Range, southern Nevada. J. Arizona Acad. Sci., 22:3-8.

————— and P. S. Martin. 1979. Ground sloth dung of the Guadalupe Mountains. Pp. 259-269, in Biological investigations in the Guadalupe Mountains National Park, Texas (H. H. Genoways and R. J. Baker, eds.). Natl. Park Serv., Proc. Trans. Ser., 4:1-442.

Stearns, C. E. 1942. A fossil marmot from New Mexico and its climatic significance. Amer. J. Sci., 240:867-878.

Stock, A. D., and W. L. Stokes. 1969. A re-evaluation of Pleistocene bighorn sheep from the Great Basin and their relationship to living members of the genus Ovis. J. Mamm., 50:805-807.

Stock, C. 1918. The Pleistocene fauna of Hawver Cave. Univ. California Publ., Bull. Dept. Geol., 10:461-515.

————. 1944. New occurrences of fossil tapir in southern California. Trans. San Diego Soc. Nat. Hist., 10:127-130.

————. 1956. Rancho La Brea, a record of Pleistocene life

in California. Los Angeles Co. Mus. Nat. Hist., Sci. Ser. 20, Paleontol. 11, 6th ed.:1-83.

⸻ and F. D. Bode. 1936. The occurrence of flints and extinct animals in pluvial deposits near Clovis, New Mexico. Part III--Geology and vertebrate paleontology of the late Quaternary near Clovis, New Mexico. Proc. Acad. Nat. Sci. Philadelphia, 88:219-241.

Stovall, J. W. 1946. A Pleistocene *Ovis canadensis* from New Mexico. J. Paleontol., 20:259-260.

Thompson, R. S., and J. I. Mead. 1982. Late Quaternary environments and biogeography in the Great Basin. Quat. Res., 17:39-55.

⸻, T. R. Van Devender, P. S. Martin, T. Foppe, and A. Long. 1980. Shasta ground sloth (*Nothrotheriops shastense* Hoffstetter) at Shelter Cave, New Mexico: Environment, diet, and extinction. Quat. Res., 14:360-376.

Van Devender, T. R., and B. L. Everitt. 1977. The latest Pleistocene and Recent vegetation of the Bishop's Cap, south-central New Mexico. Southwestern Nat., 22:337-352.

⸻ and J. E. King. 1971. Late Pleistocene vegetational records in western Arizona. J. Arizona Acad. Sci., 6:240-244.

⸻ and D. H. Riskind. 1979. Late Pleistocene and early Holocene plant remains from Hueco Tanks State Historical Park: The development of a refugium. Southwestern Nat., 24:127-140.

⸻ and W. G. Spaulding. 1979. Development of vegetation and climate in the southwestern United States. Science, 204:701-710.

⸻ and N. T. Tessman. 1975. Late Pleistocene snapping turtles (*Chelydra serpentina*) from southern Nevada. Copeia, 1975:249-253.

_____ and L. J. Toolin. 1983. Late Quaternary vegetation of the San Andres Mountains, Sierra County, New Mexico. Pp. 33-54, in The prehistory of Rhodes Canyon, survey and mitigation (P. L. Eidenbach, ed.). Human Systems Res., Inc., Tularosa, 187 pp.

_____ and F. M. Wiseman. 1977. A preliminary chronology of bioenvironmental changes during the paleoindian period in the monsoonal Southwest. Pp. 13-27, in Paleoindian life-ways (E. Johnson, ed.). West Texas Mus. Assoc., The Mus. J., 17:1-197.

_____ and R. D. Worthington. 1977. The herpetofauna of Howell's Ridge Cave and the paleoecology of the northwest-ern Chihuahuan Desert. Pp. 85-106, in Transactions of the symposium on the biological resources of the Chihuahuan Desert Region, United States and Mexico (R. H. Wauer and D. H. Riskind, eds.). Natl. Park Serv. Trans. Proc. Ser., 3:1-658.

_____, K. B. Moodie, and A. H. Harris. 1976. The desert tortoise (Gopherus agassizi) in the Pleistocene of the northern Chihuahuan Desert. Herpetologica, 32:298-304.

_____, W. G. Spaulding, and A. M. Phillips, III. 1977. Late Pleistocene plant communities in the Guadalupe Moun-tains, Culberson County, Texas. Pp. 13-30, in Biological investigations in the Guadalupe Mountains National Park, Texas (H. H. Genoways and R. J. Baker, eds.). Natl. Park Serv. Trans. Proc. Ser., 4:1-442.

Wahrhafting, C., and J. H. Birman. 1965. The Quaternary of the Pacific Mountain system in California. Pp. 299-340, in The Quaternary of the United States (H. E. Wright, Jr., and D. G. Frey, eds.). Princeton Univ. Press, Princeton, 922 pp.

Walker, D. N. 1982a. A Late Pleistocene Ovibos from southeastern Wyoming. J. Paleontol., 56:486-491.

———. 1982b. Early Holocene vertebrate fauna. Pp. 274-308, in The Agate Basin Site: A record of the paleoindian occupation of the northwestern High Plains (G. C. Frison and D. J. Stanford, eds.). Academic Press, New York, 394 pp.

——— and G. C. Frison. 1980. The Late Pleistocene mammalian fauna from the Colby Mammoth Kill Site, Wyoming. Univ. Wyoming, Contrib. Geol., 19:69-79.

Warter, J. K. 1976. Late Pleistocene plant communities--evidence from the Rancho La Brea tar pits. Spec. Publ., California Native Plant Soc., 2:32-39.

Weber, W. A. 1965. Plant geography in the Southern Rocky Mountains. Pp. 453-468, in The Quaternary of the United States (H. E. Wright, Jr., and D. G. Frey, eds.). Princeton Univ. Press, Princeton, 922 pp.

Wells, P. V. 1966. Late Pleistocene vegetation and degree of pluvial climatic change in the Chihuahuan Desert. Science, 153:970-975.

———. 1979. An equable glaciopluvial in the West: Pleniglacial evidence of increased precipitation on a gradient from the Great Basin to the Sonoran and Chihuahuan deserts. Quat. Res., 12:311-325.

——— and C. D. Jorgensen. 1964. Pleistocene wood rat middens and climatic change in the Mojave Desert: A record of juniper woodlands. Science, 143:1171-1174.

Wendorf, F. 1975. Summary and conclusions. Pp. 257-278, in Late Pleistocene environments of the Southern High Plains (F. Wendorf and J. J. Hester, eds.). Publ. Fort Burgwin Res. Cent., 9:1-290.

——— and J. J. Hester, eds. 1975. Late Pleistocene environments of the Southern High Plains. Publ. Fort Burgwin

Res. Cent., 9:1-290.

White, J. A.   1966.   A new *Peromyscus* from the Late Pleistocene of Anacapra Island, California, with notes on variation in *Peromyscus nesodytes* Wilson.   Contrib. Sci., Los
Angeles Co. Mus. Nat. Hist., 96:1-8.

Wilson, M., and C. S. Churcher.   1978.   Late Pleistocene
Camelops from the Gallelli Pit, Calgary, Alberta: Morphology and geologic setting.   Canadian J. Earth Sci., 15:
729-740.

Wilson, R. W.   1933.   Pleistocene mammalian fauna from the
Carpinteria asphalt.   Carnegie Inst. Washington Publ.,
400:59-76.

————.   1942.   Preliminary study of the fauna of Rampart
Cave, Arizona.   Caregie Inst. Washington Publ., 530:169-
185.

Wright, H. E., Jr., A. M. Bent, B. S. Hansen, and L. J.
Maher, Jr.   1973.   Present and past vegetation of the
Chuska Mountains, northwestern New Mexico.   Bull. Geol.
Soc. Amer., 84:1155-1180.

Zeimens, G., and D. N. Walker.   1974.   Bell Cave, Wyoming:
Preliminary archaeological and paleontological investiga-
tions.   Pp. 88-90, *in* Applied geology and archaeology:
The Holocene history of Wyoming (M. Wilson, ed.).   Geol.
Surv. Wyoming, Rep. Invest., 10:88-90.

Heterick Memorial Library
Ohio Northern University

DUE	RETURNED	DUE	RETURNED
1.		13.	
2.		14.	
3.		15.	
4.		16.	
5.		17.	
6.		18.	
7.		19.	
8.		20.	
9.		21.	
10.		22.	
11.		23.	
12.		24.